World Sturgeon Conservation Society – Sp
Series Editor: Harald Rosenthal (Pr
Special Publication No

© 2005 Georgii I. Ruban
ISBN 978-3-8334-4038-0

Bibliografische Information Der Deutschen Bibliothek:
Die Deutsche Bibliothek verzeichnet diese Publikation in der
Deutschen Nationalbibliografie; detaillierte bibliografische Daten
sind im Internet über <http://dnb.ddb.de> abrufbar.

Original published in 1999 in Russian by GEOS Press, Moscow, Russia, 235 pp
Translated from Russian with the assistance of Michael J. Parsley and Eliezer Gurarie
Typesetting by Bela H. Buck and Markus Geisen, Alfred Wegener Institute for Polar and Marine Research

Published and produced by Books on Demand GmbH, Norderstedt, Germany

Preface

This is the first of a series of Special Publication the World Sturgeon Conservation Society plans to publish over the next few years. The series is dedicated either to species synopsis or to specific subjects related to sturgeons, while analysing, synthesizing and summarizing the wide-spread literature on subjects such as conservation, stock enhancement, habitat rehabilitation, genetics and aquaculture (just to name a few).

Publishing scientific information related to sturgeon conservation and culture is among the aims and objectives of the Society. As outlined in the statutes, WSCS intends to act as an international forum of scientific discussion for all interested in pertinent issues on sturgeons while at the same time seeking opportunities for close co-operation at an international level.

The WSCS vision is to see stocks thriving once again in important sturgeon waters such as the Caspian and Black Seas, the North and Baltic seas, the Mediterranean Sea, rivers in China as well as in the Great Lakes region of the United States and Canada in addition to other North American watersheds. The Society therefore hopes through its activities to enhance the understanding of species protection world-wide and across national borders in regions with different cultural and political backgrounds (using the highly endangered sturgeons as case examples) in order to foster the sustainable use of natural resources that are traded globally.

The objectives of the Society are:

a. to foster the conservation of sturgeon species and restoration of stocks world-wide
b. to support the information exchange among all persons interested in sturgeons.
   This particularly applies to the following subject areas:
   - general biology
   - species and habitat protection
   - stock enhancement
   - biological resource management
   - fisheries and fisheries-related issues
   - aquaculture
   - cultural and regulatory issues
c. to promote information exchange with national, regional, international, inter-governmental organisations, educational institutions (e.g. university and scientific institutes) and non-governmental organisations (NGOs).
d. to foster and support interdisciplinary and multidisciplinary research on all aspects of sturgeons (e.g. biology, management and utilisation of sturgeons).
e. to enhance the co-operation between and among anglers, fishermen, scientists, governmental agencies, local communities and non-governmental organisations (NGOs) and international organisations.
f. to inform the public on all aspects of the status and biology of sturgeons, requirements for their effective species protection, and needs for appropriate management. The tools to achieve this objective include the use of adequate and modern publications and means of communication as well as the organisation of international conferences.

We hope this special publication finds the interest of many dealing with nature and species conservation.

Neu Wulmstorf, October, 2005

Harald Rosenthal
President, WSCS

Foreword

The book is the first monograph on the species structure and ecology of the Siberian sturgeon (*Acipenser baerii*) covering its entire natural distributional range. The book analyzes and synthesizes the results of long-term field investigations by the author on almost all large populations of the Siberian sturgeon since 1982 as well as all published data regarding this species.

The monograph contains: a) a historical review of the many investigations dealing with the taxonomic status of the Siberian sturgeon and its species structure, clarifying many of changing views on the subject over time; b) analyzes the phenotypic diversity and the overall revision of sturgeon's species structure based on data collected by the author; c) describes and analyzes the ecological forms of the species and their migrations in most of the river systems; d) presents the data on current and historic distributional ranges including maps; e) presents data on size and age composition of the different populations; f) compiles and summarizes data on growth and feeding; g) presents generalized data on maturation, time and places of spawning, fecundity, and abnormalities in reproduction in different populations; h) reports on fishery statistics since 1928, makes brief estimates and assumptions based on the limited available data on the population status in various river systems and proposes measures for their conservation; and i) analyzes the various adaptive features to different environmental conditions as they are uniquely exhibited by this sturgeon species.

Although the book compiles the most comprehensive data set ever gathered on the species, he author is aware that some serious gaps exist, in particular in the description of the natural environmental conditions (as they dominate today), the various pollutional status of the habitats, the precise stock size and other factors. These shortcomings are due to both, absence of complete information in available publications and the lack of corresponding environmental and ecological investigations complementing the specific studies on the species. The author tried seriously to gather all information available so that the overall analysis was performed to the best of present knowledge. Incorporating all the available information on the subject.

Aknowledgements

The author is particular indebted to N. V. Akimova, who conducted the histological analysis of gametogenesis in Siberian sturgeon and who collaborated in writing of the corresponding section of the book. Thanks are also due to O.V. Khalatyan, former Director of the Yakutian branch of the Eastern Siberia Fisheries Research and Designing Institute, without her active assistance the completion of this book project would not have been possible.

The author wishes to express his deep gratitude to B. V. Koshelev, L. I. Sokolov, A. I. Panaitodi, Yu. N. Pugachev, A. A. Vartanov, S. G. Gorokhov, V. G. Trushkov, A. N. Yermakov, E. P. Matveyichenko, M. A. Sivtsev, S. O. Galyuk, V. G. Kryukov, E. I. Mozhukhin, E. L. Sokolova, N. G. Spirin, V. N. Zhuravski and many others for their active and constructive participation in conducting – over many years- the field work that provided many of the data on which this book is based.

The author is also grateful to Michael J. Parsley (US Geological Survey Western Fisheries Research Center Columbia River Research Laboratory) and Eliezer Gurarie for extensive assistance in the initial translation of the text from the Russian original book into English.

The author is sincerely grateful to the scientific editor of this book series and President of WSCS, Professor Harald Rosenthal, for providing the opportunity to publish the English version in this Special Publication series, while also greatly assisting through valuable critical remarks, endless personal discussions and long-night joint editing sessions to update and streamline the text of the initial English version.

Moscow, April 2005 — Georgii Ruban

Table of contents

Preface
Foreword
Acknowledgements

Introduction		1
Materials and Methods		2

Part I:	Species Structure of the Siberian Sturgeon *Acipenser baerii* Brandt	7
1.	The history of the studies and understandings of the taxonomy of Siberian Sturgeon and its intra-specific groups	7
2.	Phenotypic diversity of the Siberian sturgeon	14
3.	The ecological forms of the Siberian sturgeon	45
4.	Conclusions	51

Part II:	Ecological characteristics of the Siberian sturgeon		55
1.	Geographical distribution of the Siberian sturgeon and it's changes under anthropogenic influences		55
2.	Physical and chemical habitat conditions and anthropogenically induced changes		60
3.	Diet and growth		82
	3.1	Diet and diet composition	82
	3.2	Growth	90
	3.3	Conclusions	102
4.	Reproduction		103
	4.1	Spawning habitats and time of reproduction	103
	4.2	Age at maturity, age and size of spawning stocks, and spawning periodicity	105
	4.3	Fecundity	111
	4.4	Features of gameto- and gonadogenesis in Siberian sturgeon	118
	4.5	Early ontogenesis	120
	4.6	Conclusions	122
5.	The state of reproductive systems in Siberian sturgeon populations and a proposed classification of reproductive disturbances potentially caused by anthropogenic impact		124
	5.1	The influence of pollution on the development and functioning of the reproductive system	124
	5.2	The reproductive system of the Siberian sturgeon in Siberian rivers	133
	5.3	A classification of morphological and other abnormalities in sturgeon reproduction	152
	5.4	Conclusion	154
6.	Stock size of Siberian sturgeon and assessment of it's populations state		155
7.	Adaptive features		161
8.	Perspectives on conservation and restoration of sturgeon populations		167
	8.1	Conservation measures for Siberian sturgeon	167
	8.2	Conservation measures for Siberian sturgeon	168
	8.3	Conservation of the Siberian sturgeon gene pool	169

References	171
Publications in Russian	171
Publications in other languages	200

THE SIBERIAN STURGEON *ACIPENSER BAERII* BRANDT

Species Structure and Ecology

by

Georgii I. Ruban
A.N. Severtsov Institute of Ecology and Evolution
Russian Academy of Sciences

Introduction

As one of the commercially highly valuable fish families, acipenserids have long attracted the attention of investigators. The Siberian sturgeon is no exception. However, in view of the wide distribution in many regions of Siberia and the difficult access to many of its remote habitats, this species was relatively poorly studied through the beginning of the 1960s, particularly in Eastern Siberia. More intensive research on the ecology of the species was conducted in the Ob and Irtysh river basins, a historical review of which can be found in the work of N. P. Votinov et al. (1975). Since 1961, a series of long-term investigations on this sturgeon was started in the lower reaches of the Lena River and this work was initiated by V. D. Lebedev, professor of ichthyology at the Moscow State University. Involved in these investigations were also L. I. Sokolov (assistant professor of the Department of Ichthyology at Moscow State University (IDMSU), V. S. Malyutin (Central Industrial-Acclimatization Department of the Ministry of Fisheries of the USSR) and N. V. Akimova (A. N. Severtsov Institute for Evolutionary Morphology and Animal Ecology (AN USSR, currently IPEE RAS), who determined the principle biological characteristics of the Lena River population which served as biological base for the subsequent inclusion of this species in the aquaculture activities regionally and (later) abroad. These scientists as well as I. I. Smolyanov (Konakovskii Industrial Experimental Branch of Research Institute of Pond Fishery) and their respective departments were also heavily involved in the development of the technical exploitation aspects of eggs extraction and transportation, while also assisting in developing incubation and rearing techniques of sturgeon throughout the life cycle (e.g. maturation) thereby building brood stocks and consistent hatchery operation. The Lena sturgeon is currently one of the principle objects of commercial sturgeon farming worldwide. It is extensively raised in Russia, Moldova, Czech Republic, Hungary, Germany, France, Italy, Korea, Chile, and several other countries.

The catastrophic collapse of most sturgeon populations, including the Siberian sturgeon, brought about by over exploitation and by man-made habitat alterations, mandates the study of their life requirements, population structure and population dynamics. The Siberian sturgeon is of particular concern since even in the 1930's it was harvested at rates comparable to contemporary catches of the anadromous species in the Caspian, Black, and Azov seas.

The development of conservation strategies for the Siberian sturgeon must be based not only on estimates of the state of separate populations, but also on the knowledge of (a) various biological, physiological and genetic aspects of the species determining the structure of stocks, (b) the multivariate adaptive features which determine the species' plasticity, and (c) the diverse conditions within their habitat range. The various factors determining species structure, including the taxonomic and population aspects, the adaptive radiation of the species, the specific adaptive features as inferred from the development of its current range, and their conditions in habitats under anthropogenic influence, have not yet been satisfactorily studied.

At the beginning of our investigations in 1982, three populations of the Siberian sturgeon were relatively well known:

- the population of the Ob and Irtysh river basins,
- those of the Lena River basin, and
- the one of the watershed of the Lake Baikal.

Data on other populations of the same species were fragmentary, making it impossible to clearly gain a full picture of the species, in particular the phenotypic diversity of the species within its distributional range, the associated differences in ontogeny, their diet and feeding habits, and last not least the reproductive mode and recruitment success. The lack of satisfactory data on the species in limits of its natural range made it also impossible to obtain reliable insights into the taxonomic structure of the species and its populations, as well as their overall ecology (including growth, feeding, reproduction, and migrations). The goal of this monograph is to fill some of these knowledge gaps by combining and comparing existing data with new findings from an area covering as much of the natural range of the species as possible.

The chapters of this monograph dealing with the species structure, morphological diversity, growth, gameto- and gonadogenesis are based on samples and measurement obtained by the author during long-term joint investigation in the waterways of Siberia carried out by the IPEE RAS (A. N. Severtsov Institute of Ecology and Evolution of Russian Academy of Science), the Yakutian branch of the Eastern Siberia Fisheries Research and Designing Institute and the Department of Ichthyology at the MSU (Moscow State University).

Populations for which we were unable to obtain own data (for example Lake Baikal, Pyasina River and others) published data were included into the analysis.

The Russian version of this monograph was published with the financial support from the Russian Foundation for Basic Research – project No 99-04-62047.

Materials and Methods

The study on phenotypic diversity, growth, diet and reproductive systems of the Siberian sturgeon *Acipenser baerii* is based on long-term sample collections undertaken between 1982 and 1995 in the lower reaches of the Ob, Yenisei, Indigirka and Kolyma rivers and throughout the Lena basin (from the delta to the lower and middle reaches and in the Aldan, a right-side tributary of the Lena River). Table 1 summarizes the data on sample areas and respective total sample sizes involved in this study.

Samples from the Ob were collected in July 1995 in the lower reaches of the river on the Kudrinskii shoals 200 km upstream from the Ob Bay. They were provided by the fishermen of the Aksarkovski' fish-processing plant using drift-net fishing with trammel nets.

Samples of Siberian sturgeon and sterlet in the Yenisei River were collected in July through September 1988-1992 in the Turukhansk region, 1,420 km from the mouth of the river (20 km upstream from the mouth of the Bakhta River). Samples for histological analyses of gonads from the Yenisei River were obtained from ice fishing

Chapter I - Materials and Methods

in November - December 1990 and generously provided by Yu. V. Mikhalyov, senior researcher of the Krasnoyarsk branch of the Eastern Siberia Fisheries Research and Designing Institute.

Samples from the lower reaches of the Lena River were collected from July through August 1986. They were obtained near the settlement of Tit-Ary, which is 152 km away from the river mouth. Samples originated from seine-catches by fishermen. Samples were also taken in several branches of the Lena delta: the Trofimovskaya (4 km upstream from Saardakh-Khaya), Olenyokskaya, (at the confluence with the Gusinaya branch at Somoyilovsky Island) and the Bykovskaya branch of the Lena River (77 km from the mouth of the branch). Sturgeon in the branches of the Lena River delta were caught using standing gill nets with 50, 75, and 100 mm mesh spacing, and by hook-and-line fishing with ledger bait.

Sturgeon samples from the middle reaches of the Lena River were collected from June to July 1984 on river sections between 1,825 and 2,282 km from the mouth of the river.

Sturgeon samples in the Aldan River were taken in 1982-1983 on sections of the river between the Petushynye Islands in the lower reaches of the river (30 km from the mouth) up to the Mamontovaya Mountain (330 km from the mouth). These samples were taken with standing gill nets of 30-90 mm mesh size (square mesh, side length) and by hook-and-line fishing with ledger bait.

Samples from the Indigirka River were collected in the lower reaches of the river in places adjacent to the commercial fishery sites located between km 157 and km 285 from the river mouth between June and August 1984, July to September in 1985 and July to August of 1987.

Table 1:
Summary of sample sizes for Siberian sturgeon and sterlet collected during the period between 1982 and 1995 according to river sections.

River	Morphometric analysis	Age determination	Histological analysis of gonads	Sampling years
Siberian sturgeon: *Acipenser baerii* Brandt				
Ob	15	12	5	1995
Yenisei, lower reaches	278	495	129	1988-1992
Yenisei, delta			33	1990
Lena, lower reaches	257	720	51	1986-1988
Lena, middle reaches	66	66	10	1984
Aldan	171	139	25	1982-1983
Indigirka	281	281	128	1984-85, 1987
Kolyma	188	191	119	1988, 1989
TOTAL:	1259	1904	500	
Sterlet *Acipenser ruthenus* Linné				
Yenisei, lower reaches	101			1988-1992

Samples from the Kolyma River were taken from July to August 1988-1989 in the middle reaches near the confluence of the left tributary, the Ozhogina River (900 km from the mouth of the Kolyma River) and in the lower reaches at 235 km (7 km downstream from the Dresvyanyi shoal) and at km 68 (at Kabachkovski Island) upstream from the river mouth.

Fishing in the Indigirka and the Kolyma rivers was performed with standing gill nets with 30, 40, 50, 70 and 100 mm mesh (square mesh, side length) and by hook-and-line (cross river set-lines with ledger-bait). During samples collection in the lower Indigirka, some fish from accidental seine catches of fishermen were also used.

All specimens underwent a complete biological analysis including total length, wet weight as well as carcass and gonad weight, sex, and maturity stage. Measurements of morphometric and meristic characters were also performed using standard methods for acipenserids (Krylova and Sokolov 1981). A vernier calliper was used on fresh samples (accuracy = 1 mm). A total of 28 morphometric traits and 6 meristic traits were determined. The morphometric traits were: total length (TL); head length (C); snout length (R), snout width (SR) and snout width at the level of barbels (SRc); distance from snout tip to base of barbels (rc), from snout tip to chondral arch of the mouth (rr), and from the base of the barbels to the chondral arch of the mouth (rl); width of mouth (So) and the width of the break of the lower lip (il); length of longest barbel (lc); horizontal diameter of the orbit (o), postorbital (po) and interorbital (io) distances; head depth at the occiput (hC) and through the center of the eyes (hCo); maximal (H) and minimal (h) body depths; antedorsal (aD), anteventral (aV), and anteanal (aA) distances; length of dorsal fin base (lD), depth of dorsal fin (hD), length of anal fin base (lA), depth of anal fin (hA), pectoral-fin length (lP), pelvic-fin length (lV), pectoventral distance (P-V), ventroanal distance (V-A). The meristic traits were: number of rays in dorsal (D) and anal (A) fins, number of dorsal (Dr), lateral (Lr), and ventral (Vr) scutes and number of gill rakers on the first arch (sp. br.).

Using morphometric characters to determine phenotypic diversity in comparative analysis encounters an array of problems associated with allometric growth and corresponding ontogenetic changes in the body proportions of the fish. This makes it necessary to compare samples within one size range only or to compare similar samples which show a similar size distribution curve. The second of these methods isn't always an option, and the first approach is incorrect, because the division of samples into size groups is always arbitrary. For these reasons, the use of Mayr's coefficient of difference *(CD)* for morphometric characteristics is impossible to employ, although the significance of the difference of the parameters in the equation of simple allometry (Ruban and Panaiotidi, 1994) can be calculated to determine the significance of the sample difference for a particular morphometric characteristic. Therefore, the morphometric analysis of phenotypic diversity of the Siberian sturgeon was performed by using multivariate methods, specifically, principle component analysis (PCA) and cluster analysis.

The validity of separating the Siberian sturgeon subspecies according to morphometric characters can be determined using Principle Component

Analysis (PCA) by comparing the level of differences reflected in the distribution of individuals in the PC space between the following three categories of samples: a) between samples from populations of the same subspecies, b) between samples from populations belonging to different subspecies, and c) between samples of two closely related species. In the latter case, a comparison of the Siberian sturgeon and the sterlet (*Acipenser ruthenus*) was performed. The PCA method using morphometric characteristics also encounters difficulties caused by allometric growth. The problem of minimizing the effect of the size of individuals arises because the main task of the study is to determine differences in form rather than in size. Solutions for analyzing shape independently of size, which can be determined by any of the initially chosen variables or any combination of selected and of the biologically relevant variables, consists of a direct study of shape as a proportions (indices), expressed as logarithms (James and McCulloch 1990). The resulting dimensionless parameters can be analyzed by one-dimensional as well as by multivariate methods (Mosimann 1970). In the PCA of the morphological traits we used logarithms of indices (derived from measurement expressed as percentages of total length). In calculating the coordinates in principal component space we used as initial variables their normative deviations. The eigen-vectors of the variance-covariance matrix were normalized to 1. The NTSYS programming package was used to perform the PCA as well as the cluster analyses.

For comparisons of samples from the Ob, Yenisei, Lena, Indigirka and Kolyma populations belonging to different subspecies, samples from the lower reaches of the rivers were used, as a valid comparison requires samples collected from similar latitudes. This follows from an observed clinal variability in the morphometric and meristic traits in the Lena basin (Ruban 1989 a and b) and the analogous meristic clines observed in the Yenisei and Ob basins (Dryagin 1947, Menshikov 1947, Podlesni 1955, Ruban and Panaiotidi 1994).

Student's paired *t*-tests were used to test the significance of differences between the means of meristic characters (Plokhinski 1970) while Mayr's coefficient of difference (CD) was used for assessing the differences of means between different samples.

Our analyses of the morphometric and meristic traits were conducted without a separation of the samples into males and females as preliminary studies have shown no significant sexual dimorphism occurs among Siberian sturgeon (Ruban 1986, 1989 a and b; Ruban and Panoiotidi 1995; Sokolov et al., 1986; Sokolov and Vasiliev 1989).

Gonad samples were fixed with Bouin's fluid for histological analysis which was conducted using generally accepted methods (Roskin and Lewinson 1957). Female gonads from early stages of maturity (stages I and II) and of males in all stages were treated stepwise with to absolute alcohol and thereafter transferred to a mixture of carbolic acid and Xylol and embedded in paraffin for sectioning; female gonads at maturity stages III and IV were treated with alcohol and chloroform (stepwise increase in concentration) and soaked in aniline oil for 5-7 days before being embedded in paraffin. Sections of 5-10 μm thickness were stained according to the Mallory's method. Microphotographs of the histological sections were taken with photomicroscope types MBI-6 and "Ergaval".

Age was determined by counting annuli rings in pectoral marginal fin ray sections (0.2 to 0.15 mm) using the interpretation methods developed by Sokolov and Akimova, (1976) for Siberian sturgeon of the Lena River.

For diet analyses, the number of stomachs taken from sturgeons was as follows: lower Indigirka River = 293 (collected between 157 and 285 km from the river mouth in 1984, 1985 and 1987); Kolyma River = 185 (collected between 235 and 900 km from the river mouth in 1988 and 1989).

Stomach contents were fairly consistent, and almost entirely composed of dark silt with mixtures of various sand grain sizes. Related to this, we undertook a determination of diet composition for a certain fraction of the sampled stomachs as identified in Table 2: 8 samples of 20 sturgeon from the Indigirka (total length from 213 to 490 mm, weight 28.8-390 g, age 1-6 years) taken 157 km from the mouth of the river, 12 samples (total length 340-625 mm, weight 100-950 g, age 3-10 years), taken 285 km from the mouth of the river, and 21 samples from the Kolyma River, 13 sample (total length 570-1,240 mm, weight 590-9,600 g, age 6-37 years) collected 235 km from the mouth, and 8 samples (total length 495-720 mm, weight 390-1,420 g, age 3-13 years) collected 900 km from the mouth. For taxonomic identification of the diet components, we used the common published guidelines for species determination by determining tables according to Chernovskii (1949), Pankratova (1970) and the *Guide to Freshwater Invertebrates* (1977).

Net heat was determined as sum of average daily temperatures in centigrade for given period of observation.

Table 2:
Number of specimens of Siberian sturgeon, collected in this study for food analyses.

Sampled rivers (distance from the mouth)	Year of sampling	Total length (TL, mm)	Body weight (g)	Age, years	n
Indigirka River 157 km	1984-1985	213-490	288-390	1-6	8
Indigirka River 285 km	1985 1987	340-625	100-950	3-10	12
Kolyma River 900 km	1988	495-720	390-1420	3-13	8
Kolyma River 235 km	1989	370-1240	590-9600	6-37	13

Part I: Species Structure of the Siberian Sturgeon *Acipenser baerii* Brandt

1. The history of the studies and understandings of the taxonomy of Siberian Sturgeon and its intra-specific groups

The taxonomic status of the various populations of the Siberian sturgeon within the natural range of the species is complex and has historically been rather complicated and intricated. In order to understand the background for confusing opinions and insights into the species population structure as well as the many fragmentary studies that did not permit to find a unifying concept quickly, the present chapter tries to analyse in detail the many approaches which led to quite some confusion but can be understood individually in their own right. With such vast geographic area coverage of a species and the relatively few and limited studies in the past, it is quite understandable that differences in opinions and misinterpretations do occur. The author uses this chapter as a unique opportunity to analyse and interpret all of the approaches and to explain the long way of gaining sufficient scientific knowledge that allow us today to present a better picture of the status of the species and intraspecific units. One of the outcomes of this review was the formulation of one of the main tasks of the work described in this book.

The studies on acipenserids in Siberian rivers began over 200 years ago. Firstly, many investigators identified not only the Siberian sturgeon but also some other sturgeon species as *Acipenser sturio*, before *Acipenser baerii* was described as a distinct species by G. F. Brandt (Brandt 1869). Secondly, it was assumed that many water bodies of Siberia outside the Ob and Yenisei watersheds are inhabited by the sterlet, *Acipenser ruthenus*. However, it became apparent later that sterlets are absent from these water bodies and that these had been misidentified and should have been identified as Siberian sturgeon. Debates regarding the presence of sterlets in the basins outside the Ob and the Yenisei watersheds continued during first half of the twentieth century.

I.G. Georgi (Georgi 1775) noted sturgeons in Lake Baikal and in the rivers that drain into it, namely the Selenga, the Angara and the Barguzin rivers, and identified the species as *Acipenser sturio* L. He also noted sterlets in the Lake Baikal and the Angara River.

Martin Sauer, on his voyage with Joseph Billings during an expedition along the Kolyma river in 1786 recorded sterlet ("stirled", Sauer 1802, p. 85) near Sredni' Ostrog (currently Srednekolymsk), but in his list of fishes in the Kolyma he cites a "sturgeon" (p. 86). In both cases, he does not identify the species by its Latin name.

Subsequent investigations on the taxonomy of acipenserids in Siberian waters were due to the work of Peter Simon Pallas. This work deserves detailed consideration. In his well-known publication (Pallas 1814), pp. 83-107 the author presents descriptions of a whole array of acipenserid species, not only giving their common names great sturgeon, stellate sturgeon, sterlet and kaluga, but also identifying their scientific name as *Acipenser huso, Acipenser helops, Acipenser pigmaeus* and *Acipenser orientalis*, respectively. Particularly noteworthy is the group described on pp. 91-97 as number 83 referring to *Acipenser sturio*. However, this may not be true and many other species may simply have been included by Pallas under this name. From synonyms cited by the author (*Acipenser shipa*), and from the local common names cited in many languages (Tatar, Yakut, Yukagir, Lithuanian, Estonian, Mordovian, Mongol, Buryat, Armenian, Persian and others) it can be inferred that under the species identified by P. S. Pallas as *Acipenser sturio* were included several species which were not described as separate species up to this time. This can also be inferred from the identified overall distribution of the group, including the Black, Azov, Caspian and Baltic seas and rivers that feed both the Arctic and Pacific Oceans, taking also into account contemporary notes on classification of Acipenseridae. Excepting just four species described by Pallas as separate species, namely the stellate, sterlet, giant sturgeon and the kaluga, the scientific name of *Acipenser sturio* formally may have stood practically for all other presently known representatives of the genus *Acipenser* in Eurasia. Pallas undoubtedly combined under the name *Acipenser sturio* the following species which were described later as separate taxonomic entities:

- the Russian sturgeon (*A. gueldenstaedti*), because Pallas' range for *A. sturio* included the Caspian sea and the Volga and Kama rivers;
- the ship sturgeon (*A. nudiventris*) – mentioned by the author as a synonym for *A. sturio* with the commentary that the sharper snouted fish are referred to in Russian as "ships";
- the Siberian sturgeon (*A. baeri*) as occurring in several Arctic Ocean Basin rivers, particularly the Yenisei and the Irtysh, were included in the known range of the species;
- the Amur sturgeon (*A. schrencki*) as rivers that flow into the Pacific ocean (orientali Oceano) are included in *A. sturio's* range; and
- the Atlantic sturgeon itself (*A. sturio*), as the Baltic and Black seas are included in the range.

Recently, however, a further distinction within the populations of the Atlantic sturgeon reveals the occurrence of two separate species in the Baltic and North Seas (Ludwig *et al.* 2003)

For a long time the Siberian sturgeon was consequently referred to as *A. sturio* by many authors working in various Siberian water bodies: at the mouth of the Lena (Figurin 1823), in Lake Baikal and the Selenga, Angara and Bagruzin rivers (Pezhemski' 1953), in the Lena, Yana, Indigirka and Kolyma rivers (Argentov 1860), in the Yenisei, Khatanga, Pyasina and Tas rivers (Tretyakov 1869), and in the Yenisei and Angara rivers (Krivoshapkin 1865).

At the same time, several authors believed that outside the Ob and Yenisei river basins the acipenserid fishes in Siberia were represented only by the sterlet (*A. ruthenus*), or both sterlet and sturgeon (*A. sturio*), or only sturgeon:

a) in the Lena River both species sterlet and sturgeon (Figurin 1823, Argentov 1860, Seroshevski' 1896) have been reported. N. A. Varpakhovski' (1898) cites two species in the Lena River, sterlet and sturgeon, but identifies sturgeon not as *A. sturio* but as *A. baeri* Brandt. Nothing but sterlet was reported in the Lena River and its tributary the Viluyi River and also in Viluyi River tributaries which are the Chona and Markha rivers by Maak (1886, 1887).

b) both sturgeon and sterlet are cited by Gedenstrom (1823) and Argentov (1860) as inhabiting the Yana River;

c) both sturgeon and sterlet are cited as inhabiting the Indigirka River (Argentov 1860);

d) In the Kolyma River, Kiber (1823) cites only sterlet, Argentov (1860) and Iokhelson (1898) cited both sturgeon and sterlet, and Rasputin (1903) mentions only the sturgeon. In publications by M. M. Gedenstrom (1823), V. I. Iokhelson (1898) and I. O. Rasputin (1903) the Latin names of the sturgeon are not given and it is scientifically uncertain which species the authors truly wished to address.

Somewhat exceptional in this regard is the work of Bunge (1883) who identified three species of ganoids in the Lena River: (a) the sterlet and (b) two other unidentified species. One of these is referred to as a "kostyor ("russisch Kostjor"), encountered in the upper reaches of the Lena River upstream from Vitimsk to the Zhigansk settlements.

In 1869, sturgeon from the Ob and the Lena rivers were described as a separate species, *Acipenser baerii* (Brandt 1869). However, even thereafter, the Siberian sturgeon appeared in publications of various authors under a variety of names: *Sturio baeri* (Dybovski, 1874 and 1876), *Acipenser gueldenstaedti* (Finsch 1876, cited by Nikolskii 1939), and *Acipenser sturio* (Bergott 1880, cited by Nikolskii 1939).

The situation became even more complicated later on: in 1896 A. M. Nikolskii, using two specimens from the lower Yenisei River (N 10885 and 10886 from the collection of the Zoological Museum of the Academy of Sciences, collected by A. Botkin on July 5, 1896), described a new species *A. stenorrhynchus* while also using two other specimens from Lake Baikal (N 10641 from the collection of the Zoological Museum of the Academy of Sciences, collected March 1, 1894 by V. Sukachov, and No 1631 from the collection of the St. Petersburg University), he described a variety as *A. stenorrhynchus var. baicalensis*. The author noted that the Baikal variety is distinguished by a shorter and wider snout and smaller pectoral fins. Later, he indicated that the ranges of *A. baeri* and *A. stenorrhynchus* coincide - both species inhabit all the major Siberian rivers that flow into the Arctic ocean and the Lake Baikal (Nikolskii 1902).

Thus, at the end of the 19th and beginning of the 20th century it was believed that Siberian waters were inhabited by three acipenserid species: *A. ruthenus*, *A. baeri* Brandt, and *A stenorrhynchus* Nikolski (Berg 1900, Varpakhovski 1902, Deryugin 1898, Lavrov 1909, Lavrov and Isachenko 1911, Ruzski, 1926). Varpakhovski (1897) identified four species of acipenserids for the Ob River although he does not provide the names of these species. However, at the same time the existence of other not yet described species in Siberia was assumed. Thus, N. Levin (1899 p. 116) writes: "With regard to the sterlet, it must be said that if the Lena species is sterlet, then it is in any case not *Acipenser ruthenus*. The Lena sterlet, if indeed it is a sterlet, will most likely be identified as a new species."

Somewhat later, V. L. Isachenko (1912) proposed for the Yenisei River the existence of only two species of acipenserids, *A. baeri* and *A. ruthenus*, and, at the same time, Isachenko following L. S. Berg (1911), considered *A. stenorrhynchus* to be a hybrid of these two species.

Subsequent investigations were focused on attributing representatives of acipenserids from various water bodies of Siberia to one of these three species of sturgeons *(A. baeri, A. ruthenus, and A. stenorrhynchus)* or their hybrids, as well as identifying their habitat range. Interesting in this respect are changes in L. S. Berg's views on the systematics and distribution of different sturgeon species in Siberia. Having originally acknowledged the existence of the sterlet, *A. ruthenus* L. and two other species of sturgeon, *A. baeri* Brandt and *A. stenorrhynchus* Nikolsky (Berg 1900) in this region, Berg considered later in 1904 *A. stenorrhynchus* as a hybrid between *A. ruthenus* and *A. baeri* (Berg 1904, 1908 a and b, 1911, 1926). His considerations at the time were very influential in forming the overall opinions in scientific circles. If he would have in 1900 cited *A. baeri* only from the Yenisei and Ob rivers and Lake Baikal, than later they all would have been identified as such for all Siberian rivers from the Ob to the Kolyma. At the same time, Berg attributed the range of sterlet to include all the rivers from the Ob to the Kolyma (Berg 1916, 1923). Isachenko (1916) was at the time of the same opinion. Essipov (1923) included both species to the Lena River fauna. Considering these similar views on the natural ranges of species the opinions on the existence of hybrids between *A. baeri* and the sterlet (Berg 1904) don't seem unreasonable. However, a publication by Berg (1916) doesn't contain any information on such hybrids. Nevertheless, in 1932 the range of sterlet and Siberian sturgeon *Acipenser baeri* was restricted by Berg to the Ob and the Yenisei basins. For the remaining rivers of Eastern Siberia, Berg offers somewhat contradictory information. On the one hand, he writes that the Lena and the Kolyma rivers are inhabited by hybrids between sterlet and Siberian sturgeon, or perhaps represent a deviating form of *A. baeri* (Berg 1932, p.59), and on the other hand he supposed *A. ruthenus* and *A. baeri* hybrids (synonym of *A. stenorrhynchus*) ranges as follows:
"... all rivers from Ob to Kolyma. In the Lena, of all the sturgeon species there is only this one hybrid, distributed from Vitim to the delta." (Berg 1932, p.59), thereby denying the presence in the Lena River of the deviating form of *A. baeri* alluded to on the same page. Furthermore, it remains unclear how a hybridization of two species can exist in the enormous area between the Lena and the Kolyma rivers in the absence of either of the parent species, which are supposedly not distributed to the east of the Yenisei River. Berg also brings forth the opinion of Dryagin that it is not a hybrid in the Kolyma, but probably a unique form of *A. baeri*. P. G. Borisov (1926, 1928 a, b) held on to a version of the hybrid nature of sturgeon form in the Lena River. He wrote (1928 a, p. 25): "... we conservatively propose to call the sturgeon captured in the Lena hybrids, despite the fact that in none of these samples and in the Lena River as a whole, as it was mentioned, have we encountered either a typical sturgeon or a typical sterlet. However, if future investigations fail to turn up a typical sturgeon or a typical sterlet in the Lena river basin, then this form will need to be considered an independent species, produced in the process of a long-term hybridization of the sturgeon with the sterlet."

Uncertainty of the taxonomic status of acipenserids in the rivers of Yakutia in the 1920's and 1930's is readily traced through a series of other investigations. Thus, P. G. Borisov (1927) names sterlet as inhabiting southern Yakutia and sturgeon (without a Latin identification) in the Kolyma River (Borisov 1929). In others publications (Averintsev 1933 a, Kossov, 1933) the sturgeon of the Lena was called "sterlet", the quotations in this case apparently indicate the authors' uncertainty in rightness of the name; "sterlet (properly hybrid between sterlet and sturgeon)" (Averintsev 1930, p.20); "khatys"(Averintsev 1933 b); "the 'sterlet' or 'khatys' in majority of cases actually hybrids between true sterlets and the Siberian sturgeon …" (Averintsev 1933 c, p. 216); "khatys", i.e. sterlet (Averintsev 1933 d, p. 263). S. V. Averinstev (1933 c) points out that in the lower Lena River besides the "khatys" one occasionally but not very often encounters typical Siberian sturgeons, and that typical sterlets can sometimes be found in the middle reaches of the river. It should be noted that the name "khatys", used by Averintsev, is a local Yakut name for sturgeons that is in use even today.

The problem of the taxonomic status of acipenserids from the Yakutian rivers at species level was finally solved by G. V. Nikolskii (1939). Having analyzed a collection of sturgeons taken by the Yenisei harvest expedition of the General Directorate of the Northern Sea Route in the Yenisei Bay in 1934-1935, he found significant variability in the majority of the morphometric characteristics of the Yenisei sturgeon related to the size of the fish. Based on P. A. Dryagin's opinion on the similarities of the Kolyma sturgeons to Ob specimens of *Acipenser baeri* in the collection of the Zoological Museum of the Academy of Sciences, and taking into account the absence of the sterlet in the Lena and the Kolyma rivers, Nikolskii considered Borisov's conjecture of the hybrid nature of the Lena River acipenserids highly unlikely. He confirmed this hypothesis by comparing the Kolyma sturgeons (according to Dryagin similar to the blunt-snouted *A. baeri* of the Ob) with Lena sturgeons and the sharp-snouted Yenisei River sturgeon. Snout length and various meristic characteristics were found similar in all three of these varieties. Based on these results, Nikolskii concluded: "… fish from the three compared water bodies are rather close in almost all traits. Some variation is observed only in the number of ventral scutes and dorsal fin rays of the Kolyma specimens. But the observed differences are so insignificant that one can hardly presume to separate the Kolyma variety into its own geographical race. On the basis of the examined materials one can quite exactly ascertain that representatives of the sturgeon family inhabiting the Yenisei, the Lena, the Kolyma and, apparently, the Ob are undoubtedly their own independent species, though, by series of characters exposed to very high individual variability. In the Ob and Yenisei, this species occurs with sterlet *A. ruthenus*. In the Lena and Kolyma, however, it is the only representative of the Acipenseridae family. With regards to hybrids with sterlet neither in the materials of P. G. Borisov nor in that of P. A. Dryagin and my own data they are apparently absent. In principle, hybridization between these two species cannot be excluded; but all of the individuals which have formerly been identified as hybrids between *A. ruthenus* and *A. baeri*, probably are not." (Nikolskii 1939, p. 146).

The hypothesis of hybrid origin of sturgeon in the rivers of Yakutia was only definitively rejected much later through the karyological analyses performed by Vasiliev *et al.* (1980).

Subsequently the problem of the taxonomic status of separate forms of the sturgeon in Siberian water bodies was examined on the subspecies level. M. I. Menshikov (1947), in comparing his own Irtysh sturgeon data with the data used earlier by Nikolskii (1939) from the Yenisei, Lena and Kolyma sturgeons, ascertained a clinal variation of meristic traits, the means of which increase from west to east. The author did not take into consideration the change in the direction of the clines between the Lena and Kolyma sturgeon, despite the significant differences in three of the five compared traits (Menshikov 1947, p. 371, Tab. 1) and the lack of significant differences between the Yenisei and Kolyma sturgeon in terms of the number of rays in the dorsal and anal fins. Thus, in this particular case, one could only speak of an increase in the traits values only from the Irtysh to the Lena River. Based on differences on the meristics and to a greater degree on morphometric traits (mainly the ratio between lengths of snout and postorbital part of the head) and based on the similarity in the morphometric traits of specimens from the Yenisei, Lena and Kolyma discovered by Nikolskii (1939), Menshikov comes to the conclusion that two forms of Siberian sturgeon exist: *Acipenser baeri* Brandt in the Ob River system and *A. baeri stenorrhynchus* A. Nikolsky in the Yenisei, Lena and Kolyma rivers. To the latter he also associated the sturgeon of the Gydanski Gulf not excluding the probability of its separation in *natio* of this subspecies. Thus, the sharp-snouted species *A. stenorrhynchus* A. Nikoslky described in the Yenisei River was reclassified as a subspecies of *A. baeri* inhabiting the Yenisei, the Lena and the Kolyma rivers.

P. A. Dryagin disagreed with a grouping into a single subspecies of the sturgeon of the Yenisei River and those of rivers in the Yakutia region, and apparently with the very separation of a subspecies called *A. baeri stenorrhynchus A.* Nikolsky (Dryagin 1947). He considered that the sturgeons in the rivers from Khatanga to Kolyma distinctly differ from the Ob sturgeon, and that the Yenisei fish displayed many intermediate characteristics. If the Ob and Yenisei sturgeon can be identified as *A. baeri* Brandt, then those of the rivers from the Khatanga to the Kolyma should be considered a subspecies called *Acipenser baeri chatys* Drjagin, or the Yakutian sturgeon. Later, Dryagin considered the separation proposed by Menshikov of the Yenisei sturgeon into a separate subspecies (Menshikov 1947) and, similarly, the separation proposed by A. G. Yegorov of a separate Baikal subspecies [cited in (Dryagin 1949)] as insufficiently supported (Dryagin 1949). Dryagin attributed the origin of *A. baeri chatys* Drjagin to the probable (following Borisov 1928) hybridization of the Siberian sturgeon with sterlet currently absent in these rivers (Dryagin 1947). The argument for this separation was supported by: (a) the long-term isolation of the Yakutian sturgeon from western Siberian sturgeon; (b) the specifics of its distribution, extending from the Khatanga River in the west to the Kolyma River in the east - a range in which other acipenserids (sterlet) are absent; and (c) the unique morphological characteristics of the Lena and Kolyma sturgeons, including the greater number of dorsal, lateral and ventral scutes and the number of gill rakers, in all of which respects the Yakutian sturgeon is more similar to that of the Amur sturgeon (Dryagin 1948 a).

The discussion whether the separation of the Yakutian sturgeon into a distinct subspecies is feasible began even before the publication of Dryagin's work. Nikolskii (1939), referring to Dryagin's manuscript and using some of Dryagin's data on the Kolyma sturgeon considered because of little morphological differences the question of considering the Yakutian sturgeon a distinct "geographical race" (subspecies by Nikolskiis meaning) can hardly be raised. L. S. Berg (1949), taking into account the community of habitat ranges of the Yakutian and eastern Siberian subspecies, concluded that the *A. baeri chatys* was synonymous to *A. baeri stenorrhynchus* A. Nik. 1896 *sensu* M. I. Menshikov (1947). A. P. Andriyashev (1954) also recognized that the insignificance of differences between the Yakutian sturgeons from the Yenisei "subspecies" made it illegitimate to separate it as a taxonomic unit higher than *natio*. N. I. Kozhin (1949) does not refer to *A. baeri chatys*, but assumes that the Siberian sturgeon living east of the Ob River is a particular form (subspecies *A. baeri stenorrhynchus*).

Subsequently for a long time the problem of Siberian sturgeon's subspecies in Yakutian water bodies remained rather intricated. Thus, *A. baeri chatys* Drjagin is attributed to the Lena River as a whole and the Indigirka rivers (Kirillov 1950 and 1955, Pirozhnikov 1959), while at the same time the east-Siberian sturgeon *A. baeri stenorrhynchus* A. Nikolsky is attributed to the river systems of Siberia from the Ob to the Kolyma River (Karantonis, Kirillov and Mukhamediyarov 1956). This same subspecies is attributed to the Viluyi River (Kirillov and Rybnikov 1958, Kirillov 1962, Lepyoshkin 1964). Later, F. N. Kirillov attributes *A. baeri chatys* Drjagin to all the rivers of Yakutia (Kirillov 1972). Yu. E. Kalashnikov (1978) considers the sturgeon of the Vitim River as *A. baeri stenorrhynchus* A. Nikolsky. Misspelling of the name *A. baeri hatys* was given instead of initial *A. baeri chatys* in a series of publications (Votinov et. al. 1975, Gundrizer et. al. 1983, Dryagin 1949, Kirillov 1950, 1955 and 1972, Pirozhnikov 1959, *The Biology of the Viluyi Reservoir* 1979, Sokolov and Vasiliev 1989), although *A. baeri chatys* was brought into initial use in 1947 by P. A. Dryagin. Also, the misspelling of the subspecies name: *A. baeri stenorhynchus* was used by several authors (Karantonis, Kirilov and Mukhamediarov 1956, Kirilov and Ribnikov 1958, Lepyoshkin 1964a, and others) instead of those names brought into use by A. M. Nikolsky for the species and later by M. I. Menshikov (1947) for the subspecies *A. baeri stenorrhynchus*.

L. I. Sokolov and V. P. Vasiliev (Sokolov and Vasiliev 1989) indicated that the taxonomic status of sturgeons in the Yakutian rivers are unclear and require further research.

Subsequent investigations (Ruban 1995a, Ruban and Panaiotidi 1994, Ruban 1994 and 1997) by one-dimensional statistical methods discovered that the morphological differences between populations in the Yenisei River (*A. baeri stenorrhyncus* as per Manshikov (1947) and in the Lena River (*A. baeri chatys* as per Dryagin (1948a) do not attain the subspecies level according to the criteria clearly defined by E. Mayr (1971) and that the sturgeons in the Yakutian waters need to be included into the subspecies of *A. stenorrhynchus*. As a result of these revisions, it was further shown that in accordance with Article 31 of the *International Code of Zoological Nomenclature* (1988) the initial spelling of the species name *Acipenser baerii* Brandt must be maintained, and that *A. baeri* (Article 33, Section d) is an incorrect subsequent spelling.

The Baikal sturgeon variety *Acipenser stenorhynchus var. baicalensis* described in 1896 by Nikolskii was reclassified by A. G. Yegorov as *A. baerii stenorrhynchus natio baicalensis* A. Nikolsky (Yegorov 1948 and 1961). Sokolov considered that the Baikal sturgeon must be separated as subspecies, *A. baeri baicalensis* (Sokolov 1966a). In subsequent publications, the opinions of authors diverged. Several scientists followed Yegorov's precedent and referred to it as *A. baerii stenorrhynchus natio baicalensis* (Kirillov 1972, Votinov et. al. 1975, Gundrizer et. al. 1983, Sokolov and Vasiliev 1989), whereas others referred to it as a subspecies *A. baerii baicalensis* (Pavlov et. al. 1994).

Thus, in the middle of the 1990's it was considered that the species *Acipenser baeri* Brandt included three subspecies: *A. baerii baerii* in the Ob system, the sharp-snouted sturgeon *A. baerii stenorrhynchus* A. Nikolsky in the Yenisei and the rivers of Eastern Siberia and *A. baerii baicalensis* in Lake Baikal.

It has to be noted that so far subspecies of Siberian sturgeon were separated on the basis of differences in morphometric and meristic characters. The difficulties described above in establishing the taxonomic status of distinct intraspecies groups of sturgeon were the result of several reasons : (a) the shortage of representative samples collected across the entire species range; (b) the lack of a unified system of morphometric measurement which makes it virtually impossible to compare the data from different authors; and (c) significant differences in the sizes of the sampled fish, which, taking into account allometric effects, makes it impossible to compare samples using morphometric traits. Furthermore, the separation of subspecies according to morphological traits was done without consistent and well-defined criteria for evaluating the level of differences in the samples according to these traits. In view of these considerations (or the lack thereof), the validity of the Siberian sturgeons subspecies separation remained doubtful.

Following the reasons pointed out above, a critical analyzes of the validity of the Siberian sturgeon subspecies separation must be based in the first place on a study of morphological differences using sufficient sample sizes and well defined criteria and statistical methods. This is the primary reason why the purpose of the following chapter is to investigate the phenotypic diversity of Siberian sturgeons with the objective to distinguish clearly the taxonomic status of the separate populations using a more comprehensive and representative sampling scheme over the entire range while also employing modern biometric and statistical methods, thereby also clarifying the validity of subspecies separations described earlier, and revealing differences between the populations in separate characters by standardized criteria.

2. Phenotypic diversity of the Siberian sturgeon

Principle component analysis on Siberian sturgeon using morphometric characters demonstrate that samples collected from the lower reaches of rivers (northern part of the range) largely – and sometimes completely – overlap. Sturgeon samples from the Yenisei River, previously attributed to *A. baerii stenorrhynchus*, and the samples from the Lena delta, previously attributed to *A. baerii chatys*, are somewhat displaced relative to each other, that is, their centers do not coincide in the space of the first two principal components, however these two samples display a high degree of overlap (Fig. 1). The samples from the Yenisei and the lower Indigirka River are even more similar (Fig. 2), as they overlap virtually completely in the space of the first two principal components, although there is some displacement of points corresponding to first and second samples. Such displacement can result due to a lack of coincidences in size composition of samples under study. Despite the measures taken to minimize the influence of specimen size it is impossible to entirely eliminate this effect. The Indigirka sample contains more of the smaller specimens than did the Yenisei one.

The samples from the Yenisei River and the lower Kolyma River exhibit more displacement than all the cases examined above (Fig. 3). Most of the Kolyma points are below the first component axis and most of the Yenisei points are above this axis. However, even with such obvious displacements, there is substantial overlap. Thus, samples from the rivers of Yakutia (Lena, Indigirka and Kolyma), classified by Dryagin as a separate subspecies *A. baerii chatys* (Dryagin 1948a), collected in lower reaches of rivers, largely or completely overlap in principal components space with samples from the lower Yenisei, which was classified as subspecies *A. baerii stenorrhynchus* (Menshikov 1947). These data agree with results obtained earlier where populations from the Yenisei and the Lena rivers were compared with one-dimensional statistical methods (Ruban and Panaiotidi 1994), showing no reason to separate the sturgeons from the rivers of Yakutia into a subspecies and the necessity of its inclusion into the subspecies *A. baerii stenorrhynchus* because the latter was described earlier than *A. baerii chatys*.

Distributions of nominative subspecies were calculated for *A. baerii baerii* samples from the lower Ob and samples of the subspecies *A. baerii stenorrhynchus* from the lower Yenisei, Lena and Indigirka rivers. The results show that the first two principal components calculated on the basis of morphometric characters displayed also overlap (Figs. 4, 5, 6). Particularly close are the samples from the Ob and the Indigirka rivers (Fig. 6). Two specimens of the Ob sample are in the middle of the Indigirka cluster of data points while others are located adjacent and to the right of the majority of data plot. Such a distribution in the principle component plots for the Ob River sample is, probably, conditioned by the size compositions: the Ob sample consists of two smaller individuals 36 and 57.5 cm in total length and 13 larger fish between 116 and 165 cm in total length, whereas the sample from the Indigirka River consists of sturgeons between 21.3 and 141 cm TL. It is obvious that with greater similarity in specimen size between the Indigirka and Ob samples, the overlap of data points would be greater.

An analogue situation exists in the Yenisei and Lena samples when compared to the Ob River samples (Figs. 4 and 5), all taken in the northern part of the range in the lower reaches of the rivers. As previously shown, smaller Ob individuals cluster in the middle of the data plots for the Yenisei and Lena sturgeon distributions, while the larger individuals tend to show up on the right panel of the principle component plot, thus indicating a high degree of overlap of Siberian sturgeon samples from northern parts of the range belonging to the nominative subspecies *A. baerii baerii* and *A. baerii stenorrhynchus* and the previously separated subspecies *A. baerii chatys*.

To assess the level of differences between populations belonging to different subspecies, it is necessary to compare differences between populations within a single subspecies. This analysis was performed on samples collected from the Lena, Indigirka and Kolyma rivers, belonging to formerly attributed *A. baerii chatys* (Dryagin 1948 a), and the Yenisei sturgeon, subsequently joined with the former into the subspecies *A. baerii stenorrhynchus* (Ruban and Panaiotidi 1994). Distribution of samples in the plot of the first two principal components from various sections of the Lena basin are presented in Figures 7 to 9 and show plots for the delta, the middle reaches (2,282 km from the mouth), and its major right tributary, the Aldan. The data points from the Lena delta and the Aldan overlap almost completely though their centers do not coincide (Fig. 7) in a similar manner as is the case in the comparison between the lower reaches of the rivers (Figs. 4-6), particularly between the Indigirka and the Yenisei (Fig. 2). The plot distributions of samples from the lower and middle reaches of the Lena are somewhat more separated from each other (Fig. 8). The plot distributions of samples from the middle Lena and the Aldan (Fig. 9) are more separated than previous ones and there is only a very slight overlap of the principal component plots. These results highlight the great morphological diversity of the fish within the Lena basin and confirm earlier conclusions on the existence of morphologically distinct populations within a single basin based on results obtained through one-dimensional statistical methods (Ruban 1989 a and b). Thus, the level of morphological differences between populations within a single basin that are not completely isolated from each other is higher than the level of differences between samples from northern parts of geographically isolated river basins (Ob, Yenisei, Lena, Indigirka and Kolyma). Even greater differences are observed between samples within formerly separated subspecies *A. baerii chatys* taken from northern part of the range - the lower reaches of Indigirka and Kolyma rivers and those taken more southerly, in the middle Lena and the Aldan rivers. If the Indigirka sample only slightly overlaps with those of the middle Lena and the Aldan (Figs. 10 and 11), than the Kolyma sample is almost entirely separate (Figs. 12 and 13). Similarly separated are those of the lower Yenisei (Figs. 14 and 15).

The distribution of Siberian sturgeon and sterlet samples in the principal component space shows that the two species, whose ranges overlap in the Ob and Yenisei basins, are clearly distinct in all the geographical regions under study (Figs. 16-22). The samples of sterlet from the Yenisei and Siberian sturgeon from the lower Ob, Yenisei, Lena, Indigirka, Kolyma and Aldan are non-overlapping. In several cases (Kolyma and Aldan rivers), they are separated by a significant distance. Interestingly, the most similar plots were found for samples of sterlet from Yenisei River and the Siberian sturgeon from middle Lena River (Fig. 22), which are significantly displaced from each other with partial overlap only. The overlap in this case is similar or greater than in the case of comparing samples from the lower reaches of the Yenisei and Kolyma rivers with samples from middle Lena and Aldan, or between samples from Aldan and middle Lena rivers (Figs. 9, 12, 13, 14, 15).

Thus, the results of the principle component analysis of morphometric characters of formerly attributed subspecies *A. baerii baerii*, *A. baerii stenorrhynchus* and *A. baerii chatys* provides evidence for a relatively low level of phenotypic diversity in the northern part of the range. Samples from southern and northern parts of the range, collected in the Lena river basin, are more distinct in the principal component space. Phenotypically distinguishable populations were discovered within this basin. The differences between these populations are higher than between geographically isolated populations in other river systems that were earlier considered as subspecies.

Having presented and discussed the distribution of samples in the principle component space we now wish to compare the percentages of the variances explained by the first two principle components. As seen in Table 3, these percentages indicate lower values for closely located parts of the range compared to those from the northern and southern ends of the range. Between Siberian sturgeon and sterlet these percentages are highest in all parts of the basin.

While studying the morphological diversity of the Siberian sturgeons we must determine which of the already used characters are the most significant for separation of samples. Analysing factor loading is the method of choice (Andreev 1980). The identification of the most significant morphometric characters is easier when the populations are well separated in the principle components. For example, in comparing the Siberian sturgeon and the sterlet, the percentage of variance explained by the first two principal components is rather high, from 72.75% to 81.42%, (Tab. 3). In all cases of separating of Siberian sturgeon samples and sterlet samples, the ninth character (the width of the break of lower lip) is most important for separating these samples, because of the value of the ninth factor loading is maximal (Tab. 4). Indeed the sturgeon and the sterlet are well differentiated according to this trait, as the width of the break of lower lip in the sturgeon is considerably greater (Berg 1949). Other morphometric characters such as snout length, width of mouth, barbel length and eye diameter are less significant for separating sturgeon and sterlet samples. From these results, it appears that in distinguishing sterlet from sturgeon by morphometric traits, head measurements are generally more significant than body proportions.

While distinguishing the most distant in the space of principal components samples of Siberian sturgeon from southern and northern parts of the range, the percentage of variance explained by the first two principal components varies widely. In the case of distinguishing the Aldan sample and samples from the northern part of the range, the largest percentage of variance explained by the first two principal components is found in the most geographically distant samples from the Ob and Kolyma rivers, with 69.75% and 69.47% of the variance explained. The least percentage of variance is explained in comparisons with the same river basin, namely the Lena delta, with 51.49% of the variance accounted for (Tab. 3). Analogously, when points from the middle Lena sample are compared, the most of the variance is explained in comparisons to the distant Ob and Kolyma, with 82.72% and 75.87%, and the least variance is explained in a comparison with the neighbouring Aldan, at 58.25%.

Values of factor loading suggest that, as for sturgeon-sterlet comparisons, traits characterizing head proportions are again more significant than body proportions at distinguishing sturgeon from the northern and southern parts of its range. To distinguish sturgeons from northern (Ob, Yenisei, Indigirka, Lower Lena and Kolyma) and southern (the Middle Lena and the Aldan) parts of its range, traits characterizing head proportions are more significant than body proportions (Tab. 4). Characters such as snout length, distance from snout tip to base of barbels, distance from snout tip to the chondral arch of the mouth, width of the break of lower lip and horizontal diameter of the orbit are the most important. The barbel length is less effective in distinguishing samples from the Aldan and the lower reaches of the Yenisei, Indigirka and Kolyma. The width of break of lower lip and the distance from snout tip to base of barbels are the most important characters for distinguishing southern samples i.e. from the Aldan and the Middle Lena (Tab. 5).

Among samples from the northern part of the range, i.e. the lower reaches of the Ob, Yenisei, Lena, Indigirka and Kolyma rivers, the percentage of variance accounted for by the first two principal components varies from 50.66% in the Ob-Yenisei case to 70.63% in the Kolyma-Lena delta case (Tab. 3).

Values of factor loading when comparing samples within the northern area only, suggest that the head proportions such as snout length, distance from snout tip to the base of barbels distance from snout tip to chondral arch of the mouth and horizontal diameter of orbit are the most effective as well as in the case of distinguishing of samples from northern and southern parts of the range. However, besides these traits the width of break of lower lip emerges as highly significant for distinguishing samples from lower reaches of Ob, delta of Lena, lower Indigirka and also from delta of Lena, lower reaches of Yenisei, Indigirka and Kolyma and samples from lower Indigirka and Kolyma. For distinguishing of Siberian sturgeon samples from northern part of the range the distance between base of barbels and chondral arch of the mouth is of lower significance (Tab. 6).

Cluster analysis was used to assess the degree of similarity of Siberian sturgeon from different parts of the range according to morphometric characters. Individuals of 40 to 80 cm TL from all of the studied samples were used in the analysis because the distribution of specimen sizes varied widely between samples. From the single-link and complete-link dendrograms shown in Figures 23 and 24, it appears that samples from the lower reaches of the Yenisei, Lena, Indigirka and Kolyma rivers form a single cluster within which the degree of similarity corresponds roughly to their geographical distance from one another. The closest similarity is displayed by samples from the neighboring and not so distant Indigirka and Kolyma river basins (Fig. 23, sample 4, 5). Samples from the Lena delta (sample 3) and the Yenisey (sample 6) joins with this cluster at lower level. Samples of sturgeons from the southern part of the range, the middle reaches of the Lena River (sample 1) and it's tributary, the Aldan (sample 2) comprise a separate cluster that links up to that of the northern part of the range at a relatively low level. As the outgroup, the sample of sterlet from the Yenesei River was used (sample 7).

The good correspondence between dendrograms constructed by different methods (single link and complete link) and their consensus tree (Fig. 25) attests to the stability of the clusters and therefore confirm the reliability of used approaches.

The results of the cluster analysis of Siberian sturgeon samples using morphometric characters generally agree with the results of the principal components analysis which demonstrated that the northern populations are most similar, despite their geographical isolation in separate and distant drainage basins. The samples from the southern part of the Lena river basin are clearly distinguished from the samples from northern parts, comprising two separate clusters in the dendrograms. The main conclusion is that within a single river basin (the Lena), phenons were distinguished between which the differences are greater than between the geographically isolated populations (e.g. the lower Yenisei, Kolyma and Indigirka rivers). The level of phenotypic diversity among sturgeon populations within the formerly acknowledged subspecies *A. baerii chatys*, inhabiting the Lena, Indigirka and Kolyma basins, is rather high and is greater than the difference between this subspecies and *A. baerii stenorrhynchus* from the Yenisei. These data confirm the conclusion, based on the results of one-dimensional analysis of the characters; on groundlessness of separating subspecies sturgeon *A. baerii chatys* from Yakutian rivers and that it is necessary to include the populations of sturgeon from these rivers in the subspecies *A. baerii stenorrhynchus* (Ruban and Panaiotidi 1994).

Unfortunately, our analysis did not include samples of sturgeon from Lake Baikal, the taxonomic status of which, as it was pointed above, remains unclear. That is why we used data from Yegorov (1961) in order to perform a cluster analysis and estimate the level of differences in morphometric traits. Mean values were calculated

Table 3:
Percentages of variance explained by the first two principle components: comparing samples of Siberian sturgeon and sterlet from various river systems (%).

River	Species						
	Siberian sturgeon						Sterlet
	Yenisei	Lena delta	Aldan	Middle Lena	Indigirka	Kolyma	Yenisei
Ob	35.63 14.93	51.87 18.02	61.08 8.66	70.07 12.65	56.49 12.14	54,39 11,33	61.35 05.08
Yenisei		47.95 15.60	48.49 11.96	46.08 17.25	44.81 13.09	35.48 17.43	61.02 12.44
Lena delta			34.50 17.99	37.48 25.70	46.73 19.75	53.84 16.79	55.35 17.39
Aldan				37.53 20.99	47.43 16.12	56.73 12.74	63.71 11.40
Middle Lena					48.37 22.76	58.64 17.23	49.23 17.43
Indigirka						53.14 12.03	58.44 20.47
Kolyma							67.42 14.07

for 14 morphometric characters of Baikal sturgeon of 40 to 80 cm in total length (the size groups used in the previous cluster analyses of our own samples). These characters include head length, head depth at the occiput; maximal and minimal body depths; antedorsal, anteventral, anteanal, pectoventral and ventroanal distances; length of the base and height of dorsal and anal fins; and

Table 4:
Factor loading obtained in pair comparisons of samples of the Siberian sturgeon (from the Ob, Yenisei, Lena, Aldan, Indigirka and Kolyma rivers) and sterlet (from the Yenisei River). (explanations for Abbreviations see listing in Chapter 1 - Material and Methods).

Character	Rivers						
	Ob	Yenisei	Lena delta	Indigirka	Kolyma	Aldan	Middle Lena
C	0.04	0.01	-0.06	-0.02	0.02	-0.08	-0,06
R	0.10	0-01	-0.16	-0.07	-0.02	-0.15	-0,18
SR	-0.08	-0.11	-0.10	-0.12	-0.08	-0.11	-0,08
SRc	-0.09	-0.12	-0.09	-0.11	-0.08	-0.11	-0,07
rc	0.25	0.15	-0.08	0.02	0.09	-0.05	-0,12
rr	0.17	0.08	-0.11	-0.03	0.04	-0.08	-0.13
rl	0.02	-0.02	-0.15	-0.10-	-0.05	-0.13	-0.13
so	-0.14	-0.15	-0.19	-0.18	-0.16	-0.23	-0.13
il	-0.92	-0.94	-0.88	-0.94	-0.97	-0.83	-0.84
lc	-0.06	-0.09	-0.15	-0.11	-0.05	-0.24	-0.20
o	0.04	-0.05	-0.17	-0.10	-0.01	-0.25	-0.26
op	-0.01	0.01	0.06	0.05	0.04	0.00	0.10
io	-0.05	-0.08	-0.04	-0.05	-0.03	-0.05	-0-03
hC	-0.08	0.02	0.01	-0.03	-0-04	0.04	0-08
hCo	-0.04	-0.02	-0.02	-0.02	-0-01	-0.06	0-03
H	-0.02	0.05	0.09	0.07	0.03	0.13	0-11
h	0.00	0.02	0.09	0.07	0.05	0.10	0.12
AD	0,00	0.01	0.02	0.02	0.01	0.04	0.04
AV	0.00	0.01	0.01	0.01	0.01	0.02	0.03
AA	0.00	0.02	0.02	0.02	0.01	0.03	0.04
LD	-0.03	0.00	-0.01	0.00	-0.03	0.02	-0.01
HD	-0.01	-0.05	-0.04	-0.03	0.00	0.05	-0.01
LA	-0.01	0.00	0.02	0.03	0.02	0.03	0.04
HA	0.04	-0.02	0.03	0.02	0.04	0.07	0.05
IP	0.03	-0.04	-0.06	-0.04	0.00	-0.07	-0.06
IV	-0.01	-0.07	-0.09	-0.08	-0.06	0.06	-0.06
P-V	-0.01	0.01	0.04	0.02	0.00	0.06	0.07
V-A	0.00	0.04	0.04	0.03	0.01	0.04	0.05

Table 5:
Factor loading obtained while pair comparison of the samples of the Siberian sturgeon from the southern (the Aldan River and middle Lena) and northern (the Ob, Yenisei, Lena, Indigirka and Kolyma rivers) parts of the range (explanations for Abbreviations see listing in Chapter 1 - Material and Methods).

Character	Aldan				
	Ob	Yenisei	Lena delta	Indigirka	Kolyma
C	-0.16	-0.17	-0.14	-0.18	-0.18
R	-0.35	-0.35	-0.38	-0.36	-0.29
SR	0.02	0.01	0.02	-0.03	-0.03
SRc	0.04	0.04	0.01	-0.02	-0.03
rc	-0.54	-0.49	-0.57	-0.48	-0.38
rr	-0.39	-0.36	-0.39	-0.35	-0.29
rl	-0.16	-0.20	-0.15	-0.17	-0.16
so	0.05	-0.07	0.09	-0.02	-0.05
il	0.38	0.35	0.40	0.42	0.53
lc	-0.13	-0.22	-0.08	-0.20	-0.25
o	-0.35	-0.38	-0.25	-0.36	-0.40
op	0.06	0.03	0.13	0.02	-0.02
io	0.03	0.05	0.03	-0.01	-0.01
hC	0.16	0.05	0.13	0.15	0.13
hCo	0.03	-0.04	0.09	-0.01	-0.05
H	0.14	0.09	0.08	0.11	0.15
h	0.08	0.13	0.07	0.07	0.08
AD	0.04	0.04	0.03	0.04	0.04
AV	0.02	0.02	0.01	0.01	0.02
AA	0.04	0.03	0.03	0.03	0.04
ID	0.09	0.06	0.04	0.07	0.10
hD	0.04	0.14	-0.01	0.05	0.04
IA	0.07	0.07	0.06	0.05	0.03
hA	0.01	0.12	0.04	0.05	0.01
IP	-0.09	-0.05	-0.02	-0.10	-0.12
IV	0.07	0.19	0.06	0.14	0.16
P-V	0.10	0.10	0.11	0.12	0.11
V-A	0.07	0.04	0.11	0.12	0.08

Character	Middle Lena					
	Ob	Yenisei	Lena delta	Indigirka	Kolyma	Aldan
C	-0.11	0.09	0.09	0.13	0.11	0.01
R	-0.29	0.28	0.26	0.32	0.23	0.18
SR	0.03	-0.04	-0.03	0.02	0.00	-0.04
SRc	0.06	-0.07	-0.01	0.00	-0.01	-0.06
rc	-0.45	0.40	0.40	0.42	0.32	0.29
rr	-0.34	0.30	0.27	0.31	0.25	0.18
rl	-0.14	0.17	0.11	0.15	0.14	0.07
so	0.06	0.00	-0.08	-0.01	0.00	-0.09
il	0.60	-0.67	-0.73	-0.62	-0.74	-0.83
lc	-0.11	0.15	0.11	0.16	0.17	0.01
o	0.30	0.29	0.22	0.30	0.31	0.14
op	0.09	-0.13	-0.11	-0.08	-0.07	-0.19
io	0.03	-0.05	-0.02	0.00	0.00	-0.03
hC	0.16	-0.08	-0.14	-0.15	-0.13	-0.10
hCo	0.06	-0.04	-0.08	-0.03	-0.02	-0.13
H	0.13	-0.05	-0.07	-0.09	-0.11	0.01
h	0.09	-0.13	-0.07	-0.08	-0.09	-0.06
AD	0.03	-0.04	-0.04	-0.04	-0.04	-0.02
AV	0.03	-0.03	-0.03	-0.02	-0.03	-0.01
AA	0.03	-0.03	-0.03	-0.03	-0.04	-0.02
ID	0.04	-0.01	-0.02	-0.04	-0.05	0.01
hD	-0.01	-0.05	-0.01	0.00	0.02	0.06
IA	0.06	-0.07	-0.05	-0.05	-0.04	-0.07
hA	-0.01	-0.08	-0.04	-0.03	0.01	0.00
IP	-0.09	0.04	0,02	0.09	0.09	0.01
IV	-0.06	0.01	-0.05	-0.02	0.02	0.17
P-V	0.08	-0.08	-0.09	-0.11	-0.09	-0.04
V-A	0.05	-0.04	-0.08	-0.11	-0.07	-0.07

length of ventral and pectoral fins.

Figure 26 represents the corresponding similarity dendrogram constructed by these 14 morphometric characters with the inclusion of the Baikal sturgeon samples from Yegorov (1961). Its structure is nearly identical to the dendrogram shown in Figure 23. The same clusters occur for Aldan and the middle Lena on one hand and Yenisei, Indigirka, Kolyma, Lower Lena on the other. Baikal sample joins the latter cluster. Interestingly, its level of similarity with samples from lower reaches of Siberian rivers is greater than the level of similarity between the middle Lena and its tributary, the Aldan River.

Decreasing the number of used characters has no effect on joining of samples from southern part of the Lena river basin (Aldan and middle Lena) and samples from the lower reaches of Siberian rivers in separate clusters. This is shown in Figs. 24 and 27, where similarity dendrograms based on complete link clustering have similar structures. The complete-link dendrogram (Fig. 27) is almost fully similar to the single-link dendrogram (Fig. 26), although in the former the level of similarity between the Baikal sturgeon and those of the lower reaches is somewhat lower than that between the middle Lena and Aldan River (e.g. Fig. 26), the last two samples forming a separate cluster.

The consensus tree (Fig. 28), combining the results of the single-link and complete-link dendrograms (Figs. 26 and 27) is identical to both and supports the robustness of the clustering structure.

The results reported above on the distribution of Siberian sturgeon samples, belonging to formerly described subspecies – nominative *A. baerii baerii*, Baicalian *A. baerii baicalensis*, east-Siberian *A. baerii stenorrhynchus* and Yakutian *A. baerii chatys* in the space of principal components, and also the results of the cluster analysis, suggest that the level of inter-population differences within one subspecies (*A. baerii chatys*) in the morphometric characters exceeds the differences between samples of different subspecies. Thus, the results of the study of phenotypic diversity structure by morphometric characteristics do not support the separation into subspecies.

Analysis of our material (Tab. 7) and published data (Isachenko 1912, Nikolskii 1939) shows that high variability of meristic traits is typical of Siberian sturgeon. These characteristics vary widely within the species distributional range, as well as within separate populations. Furthermore, their values in populations belonging to previously separated subspecies largely or completely overlap.

Mean values of meristic characters among sturgeon are significantly different between samples from different basins as well as within the limits of a single river basin (Nikolskii 1939, Menshikov 1947, Dryagin 1948 a and b, Petkevich *et al.* 1950, Petkevich 1953, Podlesnyi 1955, Yegorov 1961, Ruban 1989 a and b). See Tables 8, 9, 10 and 11.

Samples of Siberian sturgeon from different parts of the lower Kolyma River, spread over a range of 700 km, are practically identical in the means of their meristic characters (Tab. 12). Consequently, in further comparisons of Kolyma sturgeon with other populations we will use the pooled data from these two samplings.

Table 6:
Factor loading obtained while performing pair comparisonsof the samples of the Siberian sturgeon from the lower reaches of rivers. Ch = morphometric characters – explanations for abbreviations see Chapter 1 – Material and methods.

Ch	Ob				Yenisei			Lena delta		Indigirka
	Yenisei	Lena delta	Indigirka	Kolyma	Lena delta	Indigirka	Kolyma	Indigirka	Kolyma	Kolyma
C	-0.15	-0.14	-0.16	-0.16	-0.14	-0,17	-0,16	-0,16	-0,15	-0,18
R	-0,41	-0,36	-0,35	-0,38	-0,37	-0,42	-0,42	-0,37	-0,31	-0,38
SR	-0,03	0,02	0,02	-0,02	0,03	-0,06	-0,04	-0,02	-0,02	-0,08
SRc	0,05	0,04	0,04	0,00	0,06	-0,01	0,02	0,00	0,00	-0,05
rc	0,60	-0,55	-0,54	-0,57	-0,53	-0,58	-0,59	-0,51	-0,43	-0,52
rr	-0,42	-0,40	-0,39	-0,44	-0,40	-0,44	-0,44	-0,37	-0,33	-0,40
rl	-0,15	-0,17	-0,16	-0,22	-0,24	-0,23	-0,22	-0,19	-0,20	-0,22
so	0,11	0,09	0,05	0,03	0,00	-0,01	0,03	0,03	0,01	-0,01
il	0,06	0,42	0,38	0,19	0,43	0,12	0,06	0,48	0,61	0,28
lc	-0,10	-0,08	-0,13	-0,08	-0,11	-0,13	-0,10	-0,13	-0,15	-0,16
o	-0,32	-0,26	-0,35	-0,30	-0,27	-0,31	-0,31	-0,28	-0,29	-0,33
op	0,11	0,11	0,06	0,08	0,13	0,12	0,10	0,08	0,07	0,07
io	0,00	0,04	0,03	-0,01	0,06	0,02	0,01	0,01	0,01	-0,02
hC	0,13	0,15	0,16	0,08	0,04	0,04	0,04	0,13	0,12	0,09
hCo	0,03	0,08	0,03	0,02	0,02	0,00	0,00	0,03	0,01	-0,01
H	0,05	0,12	0,14	0,20	0,08	0,08	0,11	0,10	0,13	0,15
h	0,05	0,08	0,08	0,08	0,13	0,10	0,08	0,08	0,08	0,07
AD	0,03	0,03	0,04	0,04	0,02	0,03	0,03	0,02	0,02	0,03
AV	0,02	0,02	0,02	0,02	0,00	0,01	0,01	0,00	0,01	0,01
AA	0,03	0,03	0,04	0,04	0,02	0,03	0,03	0,03	0,03	0,04
lD	0,08	0,04	0,09	0,07	0,01	0,06	0,06	0,04	0,06	0,08
hD	-0,10	-0,04	0,04	-0,09	0,00	-0,03	-0,08	-0,04	-0,07	-0,08
lA	0,09	0,05	0,07	0,08	0,06	0,09	0,09	0,04	0,02	0,06
hA	-0,11	0,00	0,01	-0,07	0,06	0,02	-0,07	0,01	-0,03	-0,04
lP	-0,12	-0,07	-0,09	-0,12	-0,04	-0,09	-0,10	-0,08	-0,10	-0,13
lV	-0,09	-0,01	0,07	-0,06	-0,02	-0,02	-0,06	0,02	-0,02	-0,02
P-V	0,10	0,10	0,10	0,12	0,08	0,11	0,12	0,11	0,09	0,12
V-A	0,09	0,08	0,07	0,11	0,06	0,11	0,11	0,12	0,09	0,14

The meristic traits of the Siberian sturgeon from the Lena River basin display a clinal variability (Tab. 11). In samples collected in the northern end of the range (the Lena delta), all of the meristic means, with the exception of the number of gill rakers, are greater than in the southern parts of the range (the middle reaches of the Lena and the Aldan rivers). In the Aldan samples, these means are likewise higher than in the middle Lena, taken further south. These mean differences are significant for such traits as the number of rays in the dorsal and anal fins, the number of scutes in the dorsal row and between samples from all three areas, differences in the number of scutes in the ventral row are significant between samples from delta of Lena and middle Lena and between the ones from the delta of Lena and from the Aldan (Tab. 13).

It was previously shown (Ruban and Sokolov 1986, Ruban 1989 a and b, Ruban 1992) that clinal variability of Siberian sturgeon by meristic characteristics is correlated with temperature in the respective sections of the Lena basin. It has been experimentally shown that the rearing

Table 7:
Limits of meristic characters variation in various Siberian sturgeon populations (our data). (explanations for abbreviations see listing in Chapter 1 - Material and Methods).

Characters	Subspecies						
	A. baerii baerii	A. baerii stenorrhynchus	A. baerii chatys				
	Lower Ob	Lower Yenisei	Lena delta	Aldan	Middle Lena	Lower Indigirka	Lower Kolyma
D	37-49	33-58	35-53	35-51	34-53	35-54	32-51
A	22-28	20-32	21-33	18-28	15-28	20-31	21-37
Dr	12-16	11-19	12-18	11-17	11-16	12-20	12-19
Lr	37-49	35-60	37-56	37-58	39-53	38-53	35-56
Vr	8-11	8-14	8-13	8-13	8-12	8-14	8-15
sp.br.	24-37	25-44	28-48	29-46	27-42	22-48	27-44

Table 8:
Mean values of meristic characters of the Siberian sturgeon from various parts of the Ob River basin (explanations for abbreviations see listing in Chapter 1 - Material and Methods).

Characters	Taz Bay (Dryagin 1948 b)	Ob near Belogorye (Dryagin 1948 b)	Middle Ob (Petkevich 1953)	Upper Ob (Petkevich et al. 1950), n=48-50	Irtysh (Menshikov 1947), n =60-100	P
	1	2	3	4	5	4-5
D	42.70	-	42.9	43.49 ± 0.60	42.58 ± 0.35	>0.05
A	24.70	-	23.80	23.44 ± 0.21	24.63 ± 0.31	<0.01
Dr	14.40	14.10	13.70	13.32 ± 0.14	13.52 ± 0.11	>0.05
Lr	44.10	44.10	43.40	43.64 ± 0.34	41.42 ± 0.38	<0.001
Vr	10.1	10.5	10.10	10.62 ± 0.32	9.51 ± 0.10	>0.05
sp.br	32.8	-	32.0	-	30.38 ± 0.39	-

Table 9:
Mean values of meristic characters of the Siberian sturgeon from various parts of the Yenisei River (explanations of abbreviations see listing in Chapter 1 - Material and Methods).

Characters	Yenisei Bay (Nikolskii 1939), n=43	Yenisei delta (Podlesnyi 1955), n=60	3) Yenisei between Yartsevo and Vorogovo (Podlesnyi 1955), n=139	P		
	1	2	3	1-2	1-3	2-3
D	45.2 ± 0.50	44.7 ± 0.31	43.9 ± 0.30	>0.05	>0.05	>0.05
A	24.7 ± 0.40	24.8 ± 0.22	23.2 ± 0.30	>0.05	<0.01	<0.001
Dr	14.7 ± 0.30	14.4 ± 0.18	14.0 ± 0.10	>0.05	<0.05	>0.05
Lr	46.6 ± 0.60	47.2 ± 0.57	48.6 ± 0.30	>0.05	<0.01	<0.05
Vr	10.6 ± 0.20	10.8 ± 0.13	10.7 ± 0.10	>0.05	>0.05	>0.05
sp.br.	33.60 ± 0.40	36.3 ± 0.44	36.02 ± 0.23	<0.001	<0.01	>0.05

Table 10:
Mean values of meristic characters of the Siberian sturgeon from the Lake Baikal (Yegorov 1961, explanations for Abbreviations see listing in Chapter 1 - Material and Methods).

Characters	Means	n
D	44.07 ± 0.28	136
A	25.51 ± 0.19	136
Dr	15.07 ± 0.08	285
Lr	50.06 ± 0.19	474
Vr	12.01 ± 0.06	425
sp.br.	35.04 ± 0.32	154

of the sturgeon in warm-water aquaculture is accompanied by a significant decrease in the mean values of these meristic traits compared to the values of the lower Lena population from which the sturgeon were initially taken.

These data are useful for explaining the emergence of these clines. Changes in meristic characters for warmwater-cultivated sturgeon can be explained as a result of modification variability, i.e. the change of the creod within the norms of reaction. The appearance under natural conditions of clines parallel to modification changes obtained under experimental conditions can be explained in two ways. First, we can't exclude that the observed clines may result from modification variability. Second, the clines may be a manifestations of the Baldwin's effect, that is, the fixation of what was initially a modification response through stabilizing selection (Schmalhausen 1968, Waddington 1957). In this second case, clines are manifestations of plasticity – changes in the creod, accompanying changes in reaction norms under the influence of stabilizing selection (Ruban 1992). The mechanisms behind the formation of these clines are, however, not particularly important in the study of species structure and phenotypic diversity. More important is the fact

Table 11:
Mean values of meristic characters of the Siberian sturgeon from different populations (our data, abbreviations of characters as explained in Chapter 1 Material and Methods).

Characters	Lower Ob n=15	Lower Yenisei n=278	Lena delta n=259	Aldan n=171
D	41.60 ± 1.03	44.12 ± 0.20	42.87 ± 0.21	41.25 ± 0.22
A	24.53 ± 0.48	24.69 ± 0.12	25.22 ± 0.13	22.55 ± 0.15
Dr	13.87 ± 0.26	14.82 ± 0.08	14.92 ± 0.07	14.16 ± 0.09
Lr	44.00 ± 0.86	47.97 ± 0.25	45.50 ± 0.23	45.51 ± 0.28
Vr	9.93 ± 0.28	10.88 ± 0.07	10.33 ± 0.06	10.20 ± 0.07
sp.br	30.73 ± 0.96	35.47 ± 0.22	36.69 ± 0.21	37.11 ± 0.25

Characters	Middle Lena n=66	Lower Indigirka n=281	Lower Kolyma n=188
D	40.09 ± 0.50	42.80 ± 0.18	43.01 ± 0.21
A	20.20 ± 0.30	25.48 ± 0.11	24.71 ± 0.14
Dr	13.47 ± 0.15	14.94 ± 0.08	15.08 ± 0.10
Lr	45.12 ± 0.39	45.69 ± 0.18	45.17 ± 0.25
Vr	10.08 ± 0.10	10.22 ± 0.07	10.15 ± 0.10
sp.br	35.23 ± 0.38	34.62 ± 0.21	35.52 ± 0.23

Table 12:
Mean values of meristic characters of the Siberian sturgeon from different sections of the Kolyma Ruver (a - 900 km from the mouth; b - 235 km from the mouth. Our data, explanations for abbreviations see listings in Chapter 1 - Material and Methods).

Section of the river Characters	a n = 94	b n = 97	P
D	43.09±0.31	42.94±0.30	>0.05
A	24.73±0.22	24.69±0.18	>0.05
Dr	15.05±0.14	15.12±0.13	>0.05
Lr	45.10±0.38	45.24±0.31	>0.05
Vr	10.18±0.14	10.13±0.13	>0.05
sp.br.	35.78±0.33	35.26±0.32	>0.05

Table 13:
Significance (P) of differences in mean values of meristic characters (Table 11) between various samples of the Siberian sturgeon (t-criteria). Abbreviations for characters are explained in Material and Methods (Chapter 1).

River	Yenisei	Lena delta	Aldan	Middle Lena	Indigirka	Kolyma	Characters
Ob	<0,05	>0,05	>0,05	>0,05	>0,05	>0,05	D
	>0,05	>0,05	<0,001	<0,001	>0,05	>0,05	A
	<0,001	<0,001	>0,05	>0,05	<0,001	<0,001	Dr
	<0,001	>0,05	>0,05	>0,05	>0,05	>0,05	Lr
	<0,001	>0,05	>0,05	>0,05	>0,05	>0,05	Vr
	<0,001	<0,001	<0,001	<0,001	<0,001	<0,001	sp.br.
Yenisei		<0,001	<0,001	<0,001	<0,001	<0,001	D
		<0,01	<0,001	<0,001	<0,001	>0,05	A
		>0,05	<0,001	<0,001	>0,05	<0,05	Dr
		<0,001	<0,001	<0,001	<0,001	<0,001	Lr
		<0,001	<0,001	<0,001	<0,001	<0,001	Vr
		<0,001	<0,001	>0,05	<0,01	>0,05	sp.br.
Lena delta			<0,001	<0,001	>0,05	>0,05	D
			<0,001	<0,001	>0,05	<0,01	A
			<0,001	<0,001	>0,05	>0,05	Dr
			>0,05	>0,05	>0,05	>0,05	Lr
			>0,05	<0,05	>0,05	>0,05	Vr
			>0,05	<0,001	<0,001	<0,001	sp.br.
Aldan				<0,05	<0,001	<0,001	D
				<0,001	<0,001	<0,001	A
				<0,001	<0,001	<0,001	Dr
				>0,05	>0,05	>0,05	Lr
				>0,05	>0,05	>0,05	Vr
				<0,001	<0,001	<0,001	sp.br.
Middle Lena					<0,001	<0,001	D
					<0,001	<0,001	A
					<0,001	<0,001	Dr
					>0,05	>0,05	Lr
					>0,05	>0,05	Vr
					>0,05	>0,05	sp.br.
Indigirka						>0,05	D
						<0,001	A
						>0,05	Dr
						>0,05	Lr
						>0,05	Vr
						<0,01	sp.br.

that morphological differences between samples from different parts of the Lena river basin together with data on existing spawning sites in different parts of the Lena River basin assert the existence of locally distinct populations. These spawning sites exist near the mouth of Natara river and probably the lower up to upper section of the Lena delta (Pirozhnikov 1955), as well as in the middle reaches of the Lena (Koshelev et al. 1989) and in the Aldan River (Sokolov et al. 1986). At these spawning sites the spawning times differ as shown by histological analysis of gonads. These data make it possible to assert that within the limits of the Lena basin there exist local, morphologically distinct populations of the sturgeon (Ruban 1989 a, Koshelev et al. 1989). This conclusion agrees

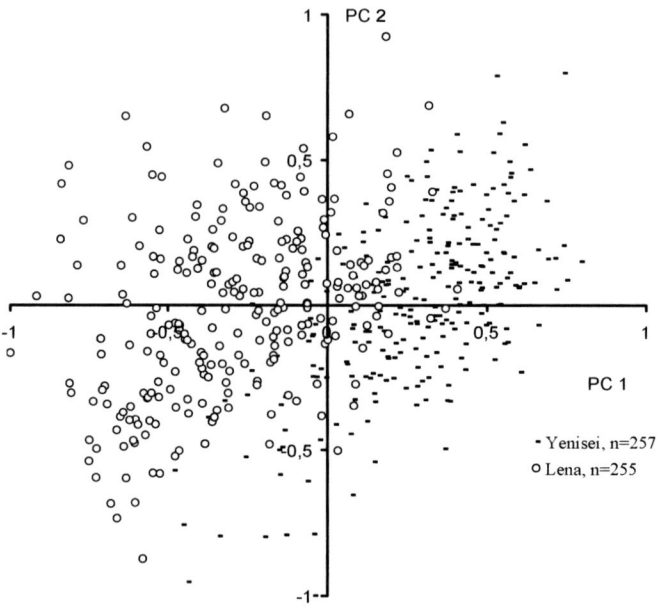

Fig. 1. Distribution of the Siberian sturgeon samples from Yenisei River and delta of the Lena River in the space of principal components (morphometric characters).

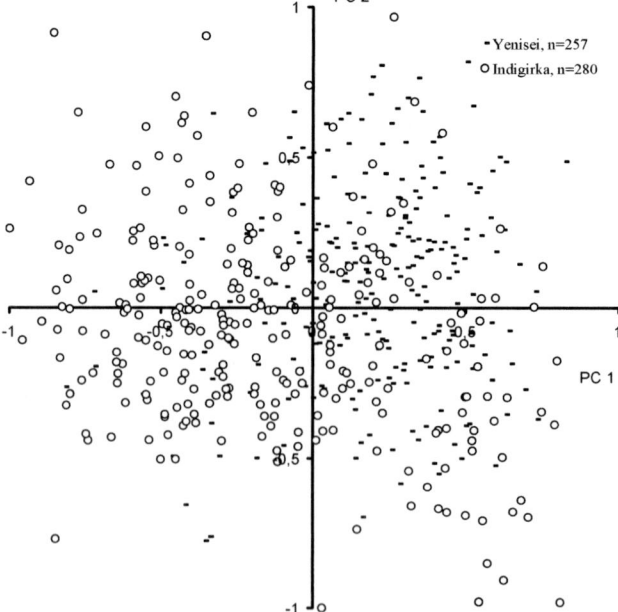

Fig. 2. Distribution of the Siberian sturgeon samples from Yenisei and Indigirka rivers in the space of principal components (morphometric characters).

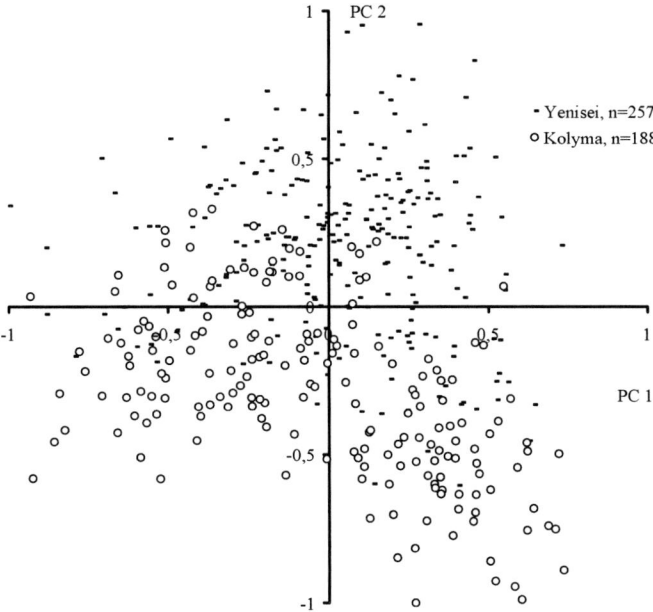

Fig. 3. Distribution of the Siberian sturgeon samples from Yenisei and Kolyma rivers in the space of principal components (morphometric characters).

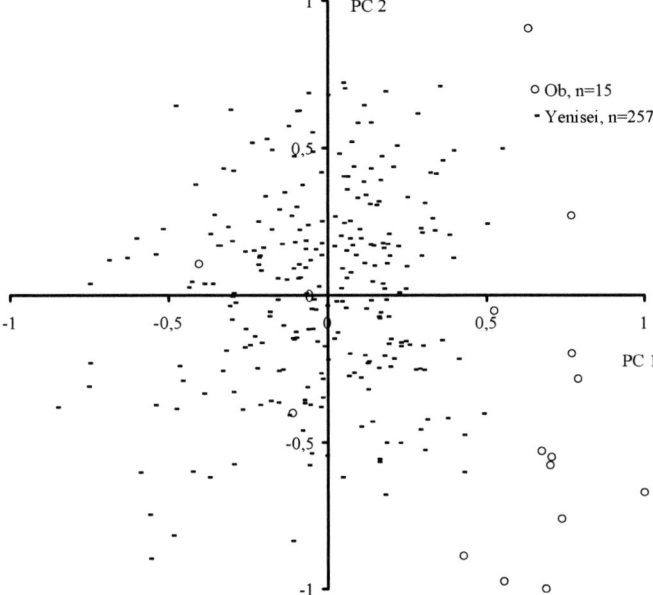

Fig. 4. Distribution of the Siberian sturgeon samples from Ob and Yenisei rivers in the space of principal components (morphometric characters).

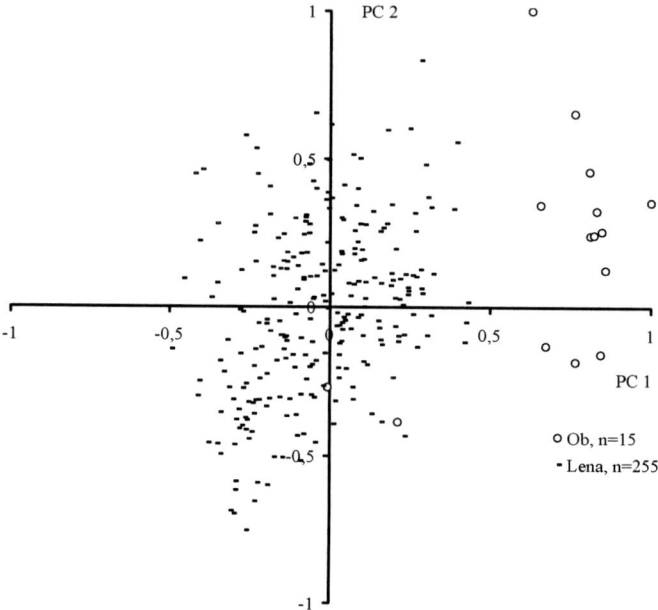

Fig. 5. Distribution of the Siberian sturgeon samples from Ob River and delta of the Lena River in the space of principal components (morphometric characters).

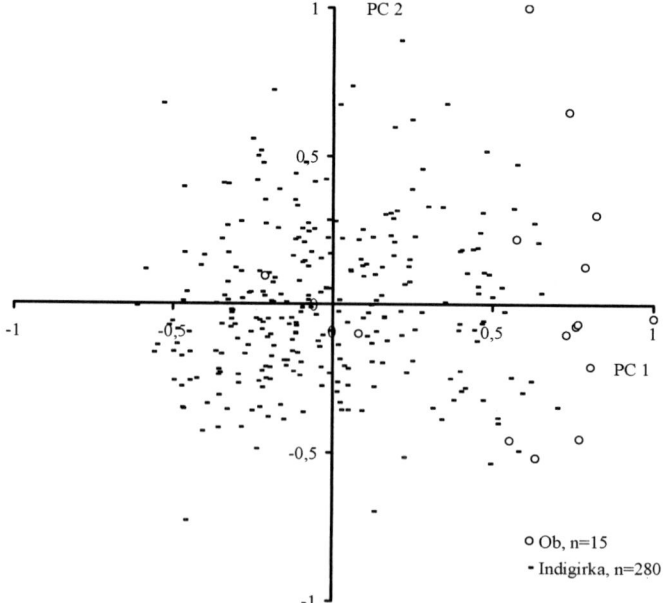

Fig. 6. Distribution of the Siberian sturgeon samples from Ob and Indigirka rivers in the space of principal components (morphometric characters).

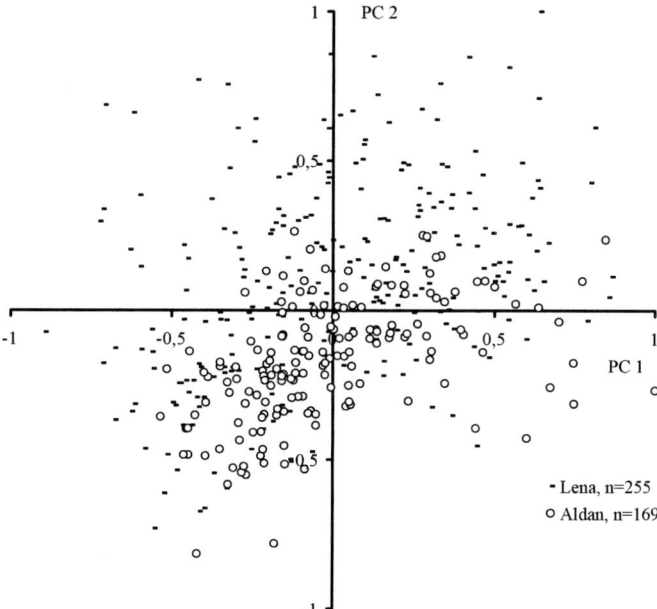

Fig. 7. Distribution of the Siberian sturgeon samples from delta of the Lena River and from Aldan River in the space of principal components (morphometric characters).

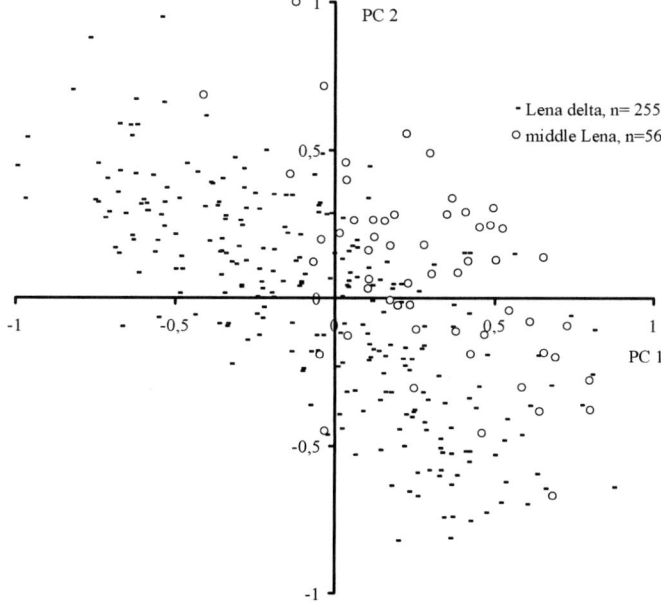

Fig. 8. Distribution of the Siberian sturgeon samples from delta and middle reaches of the Lena River in the space of principal components (morphometric characters).

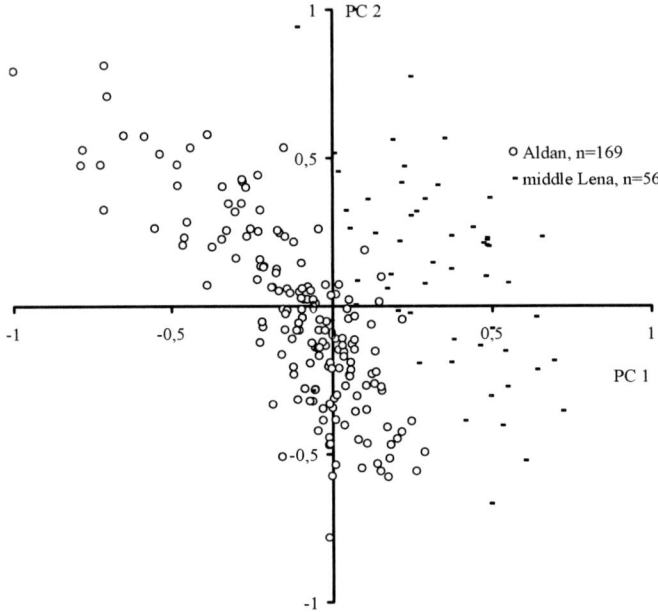

Fig. 9. Distribution of the Siberian sturgeon samples from the Aldan River and middle reaches of the Lena River in the space of principal components (morphometric characters).

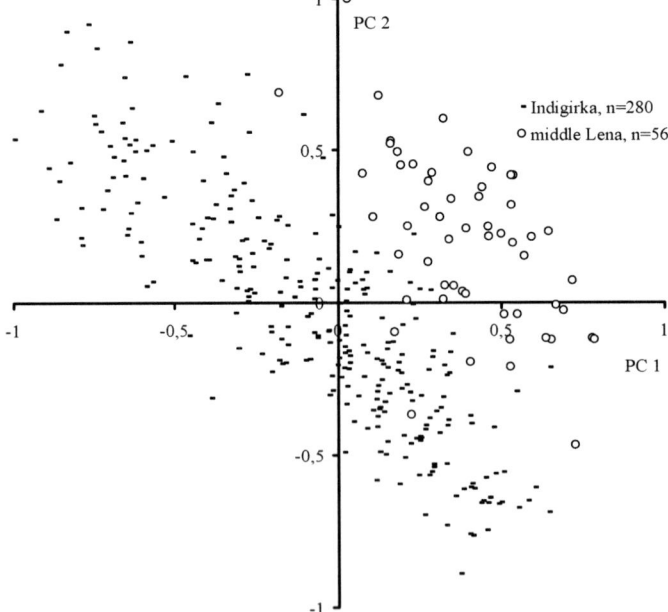

Fig. 10. Distribution of the Siberian sturgeon samples from the Indigirka River and middle reaches of the Lena River in the space of principal components (morphometric characters).

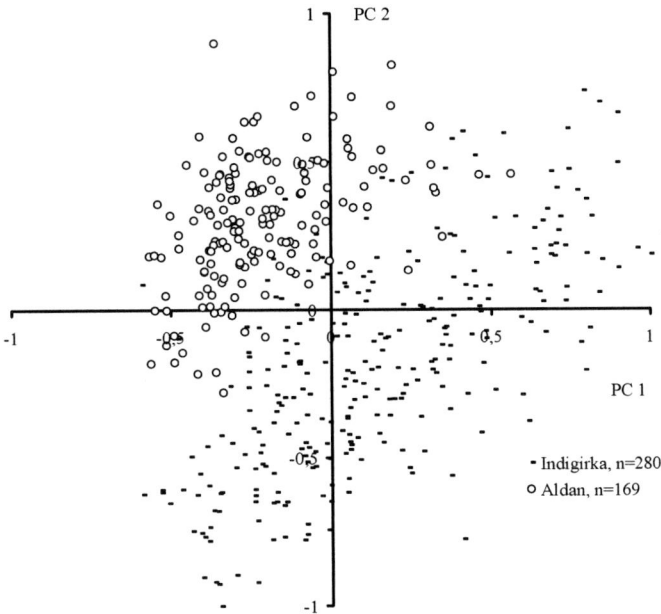

Fig. 11. Distribution of the Siberian sturgeon samples from the Indigirka River and the Aldan in the space of principal components (morphometric characters).

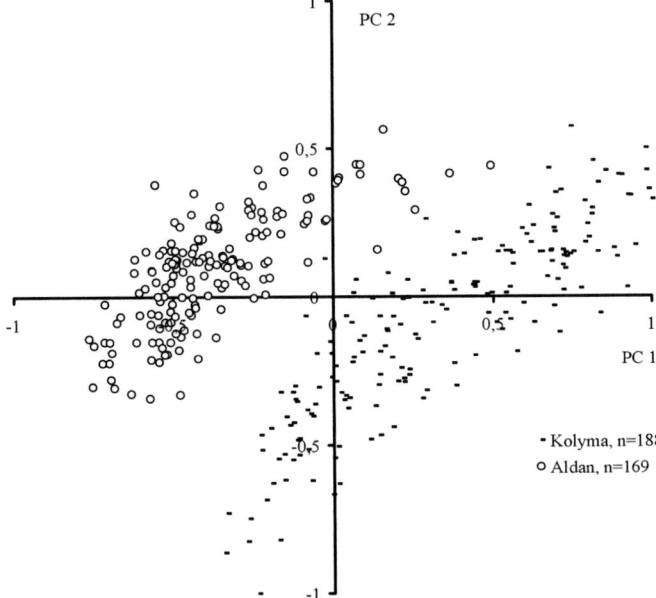

Fig. 12. Distribution of the Siberian sturgeon samples from the Kolyma River and the Aldan River in the space of principal components (morphometric characters).

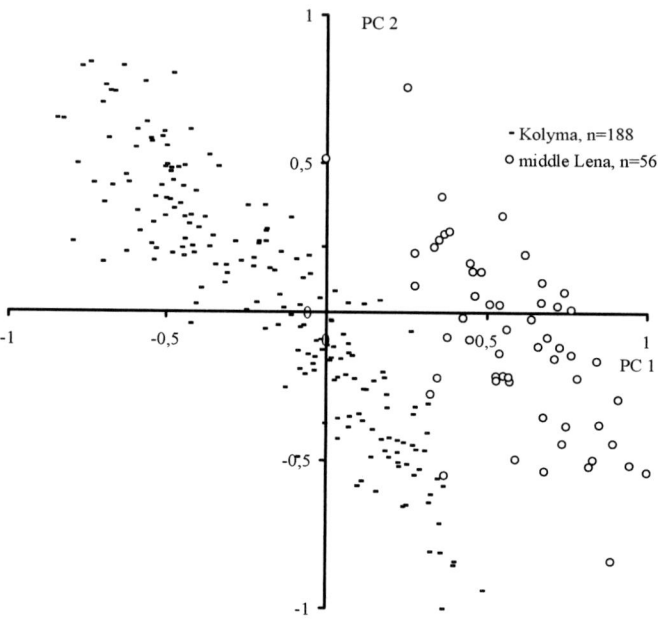

Fig. 13. Distribution of the Siberian sturgeon samples from the Kolyma River and middle reaches of the Lena River in the space of principal components (morphometric characters).

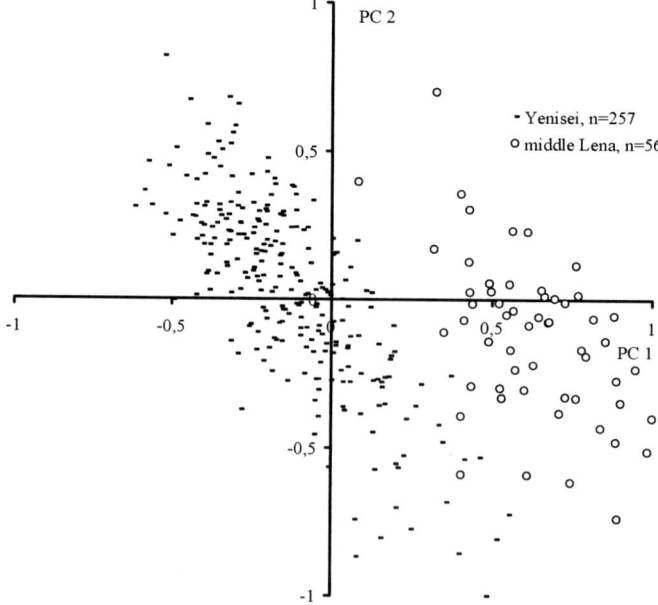

Fig. 14. Distribution of the Siberian sturgeon samples from the Yenisei River and middle reaches of the Lena River in the space of principal components (morphometric characters).

Part I - Species Structure of the Siberian Sturgeon *Acipenser baerii* Brandt

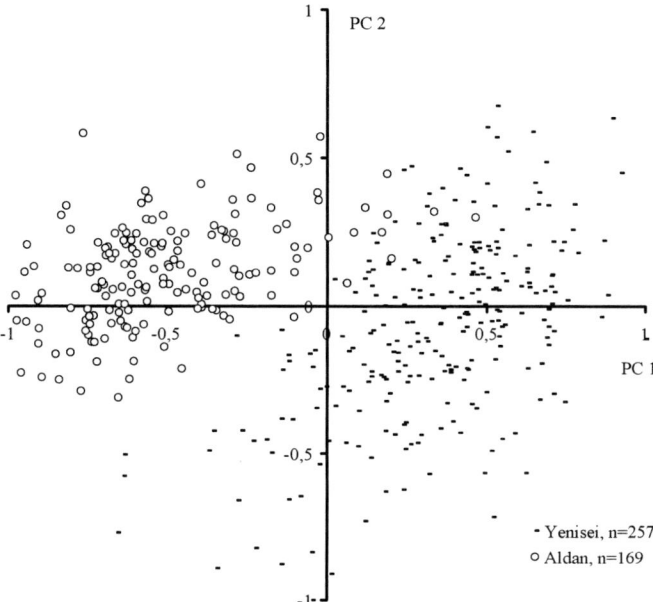

Fig. 15. Distribution of the Siberian sturgeon samples from the Yenisei River and the Aldan River in the space of principal components (morphometric characters).

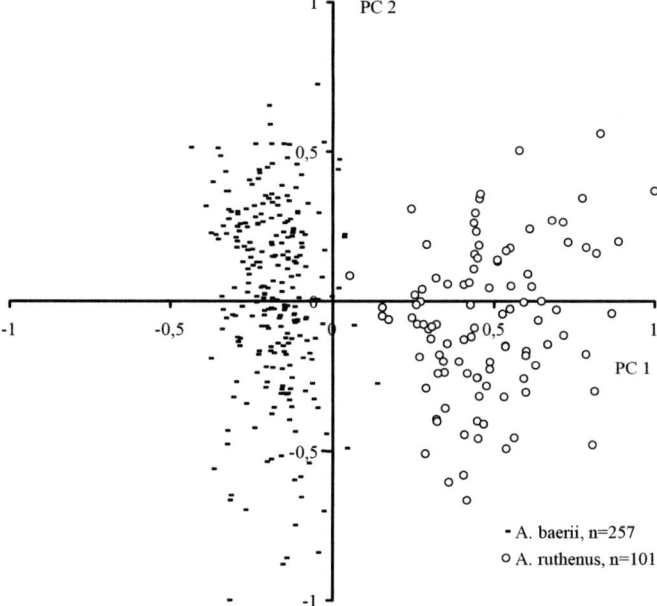

Fig. 16. Distribution of the Siberian sturgeon and sterlet sturgeon samples from the Yenisei River in the space of principal components (morphometric characters).

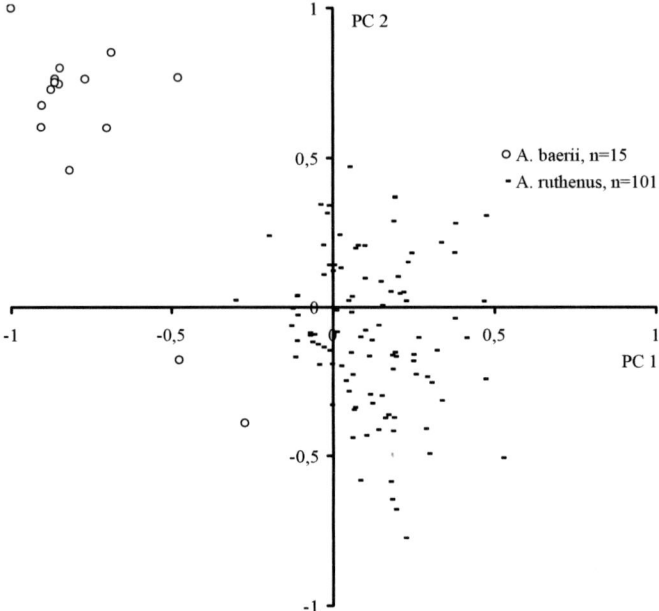

Fig. 17. Distribution of samples of the Siberian sturgeon from the Ob River and sterlet sturgeon from the Yenisei River in the space of principal components (morphometric characters).

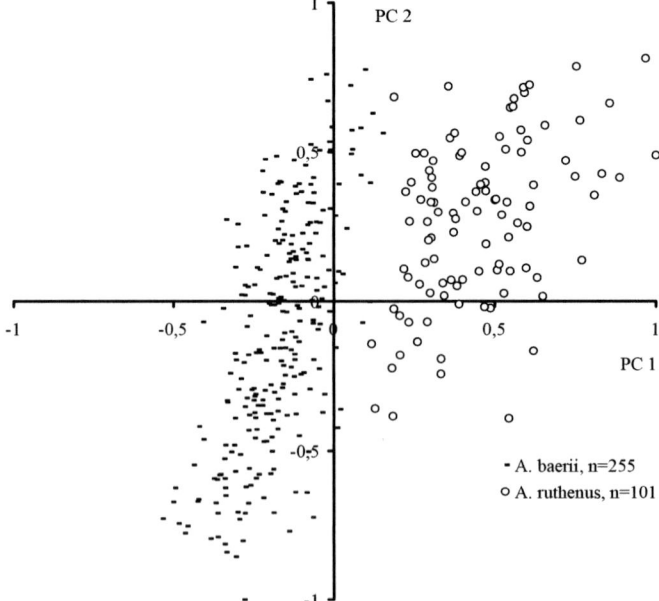

Fig. 18. Distribution of samples of the Siberian sturgeon from the Lena River and sterlet sturgeon from the Yenisei River in the space of principal components (morphometric characters).

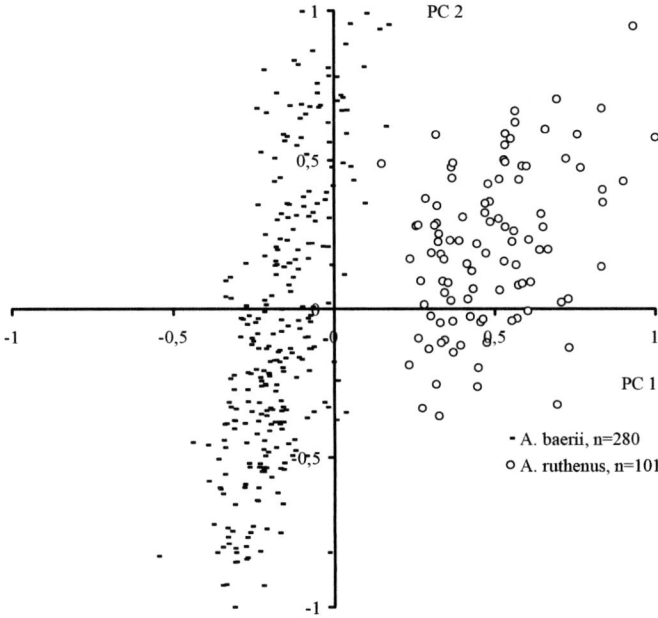

Fig. 19. Distribution of samples of the Siberian sturgeon from the lower reaches of the Indigirka River and sterlet sturgeon from the Yenisei River in the space of principal components (morphometric characters).

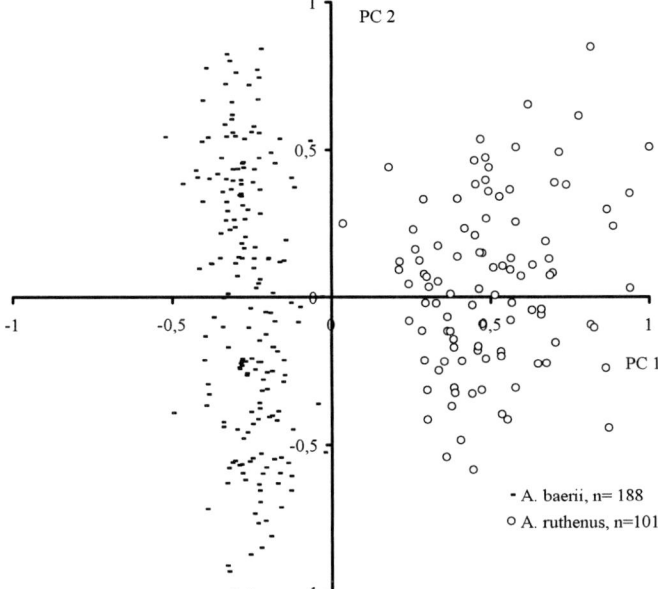

Fig. 20. Distribution of samples of the Siberian sturgeon from the lower reaches of the Indigirka River and sterlet sturgeon from the Yenisei River in the space of principal components (morphometric characters).

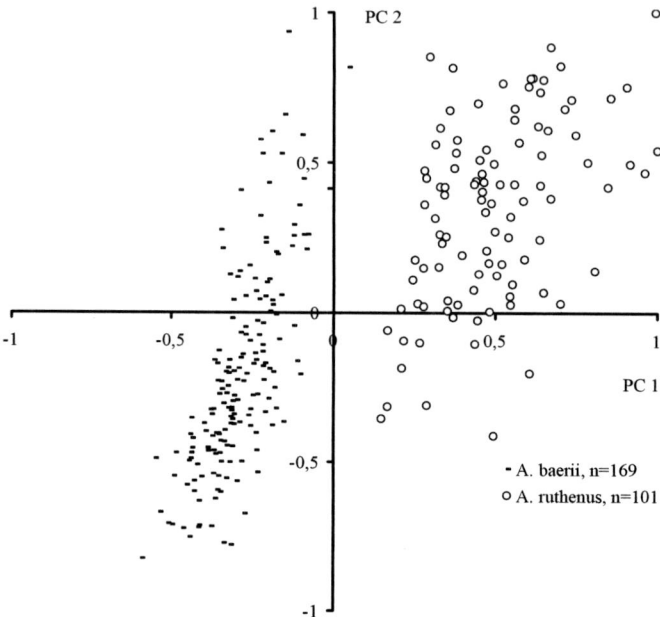

Fig. 21. Distribution of samples of the Siberian sturgeon from the Aldan River and sterlet sturgeon from the Yenisei River in the space of principal components (morphometric characters).

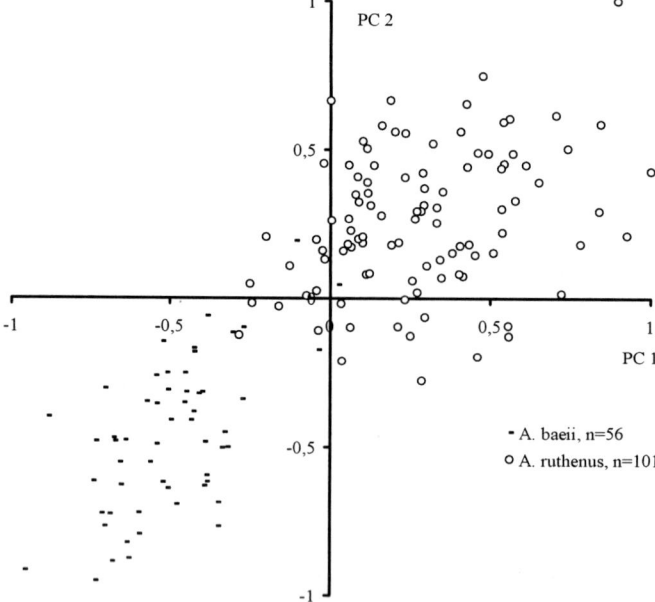

Fig. 22. Distribution of samples of the Siberian sturgeon from middle reaches of the Lena River and sterlet sturgeon from the Yenisei River in the space of principal components (morphometric characters).

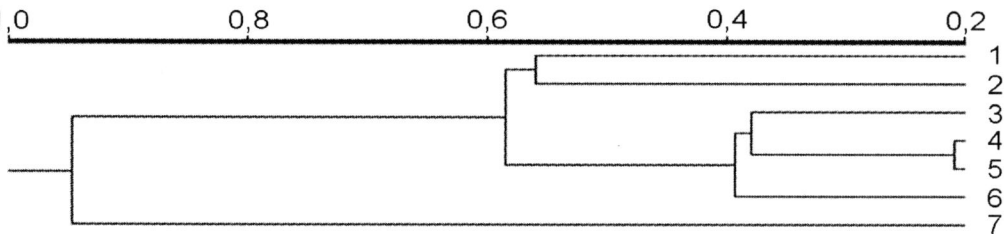

Fig. 23. The single-link dendrogram of the Siberian sturgeon samples (1 – middle reaches of the Lena River, 2 – Aldan River, 3 – delta of Lena River, 4 – lower reaches of Indigirka River, 5 - lower reaches of Kolyma River, 6 - lower reaches of Yenisei River) and sample of sterlet sturgeon from lower reaches of Yenisei River (7) based on morphometric characters.
Fig. 24. The complete-link dendrogram of the Siberian sturgeon samples (1 – middle reaches of the Lena

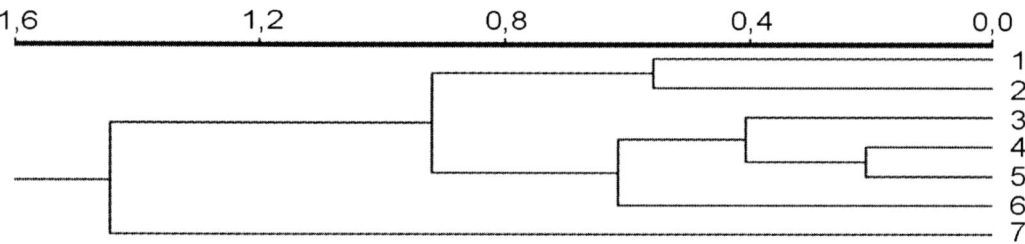

River, 2 – Aldan River, 3 – delta of Lena River, 4 – lower reaches of Indigirka River, 5 - lower reaches of Kolyma River, 6 - lower reaches of Yenisei River) and sample of sterlet sturgeon from lower reaches of Yenisei River (7) based on morphometric characters.
Fig. 25. The consensus tree of the Siberian sturgeon samples (1 – middle reaches of the Lena River, 2

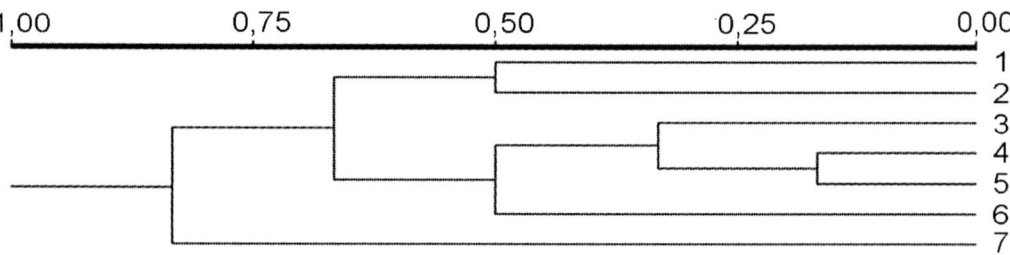

– Aldan River, 3 – delta of Lena River, 4 – lower reaches of Indigirka River, 5 - lower reaches of Kolyma River, 6 - lower reaches of Yenisei River) and sample of sterlet sturgeon from lower reaches of Yenisei River (7) based on morphometric characters, constructed using single-link and complete-link dendrograms (fig. 23, 24).
Fig. 26. The single-link dendrogram of the Siberian sturgeon samples (1 – middle reaches of the Lena River, 2 – Aldan River, 3 – delta of Lena River, 4 – lower reaches of Indigirka River, 5 - lower reaches of

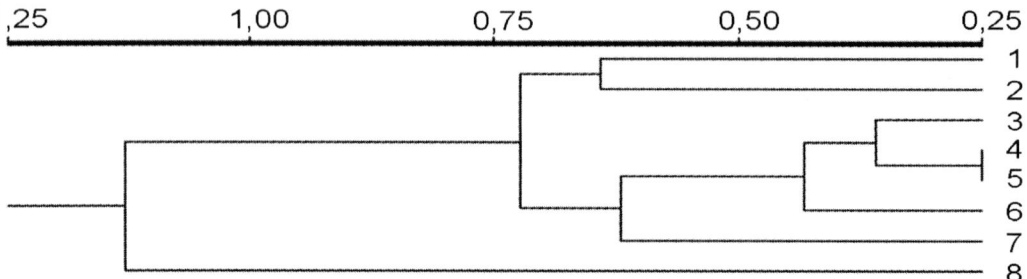

Kolyma River, 6 - lower reaches of Yenisei River), 7 – Lake Baikal (data by Yegorov, 1961) and sample of sterlet sturgeon from lower reaches of Yenisei River (8) based on morphometric characters.
Fig. 27. The complete-link dendrogram of the Siberian sturgeon samples (1 – middle reaches of the Lena River, 2 – Aldan River, 3 – delta of Lena River, 4 – lower reaches of Indigirka River, 5 - lower reaches of

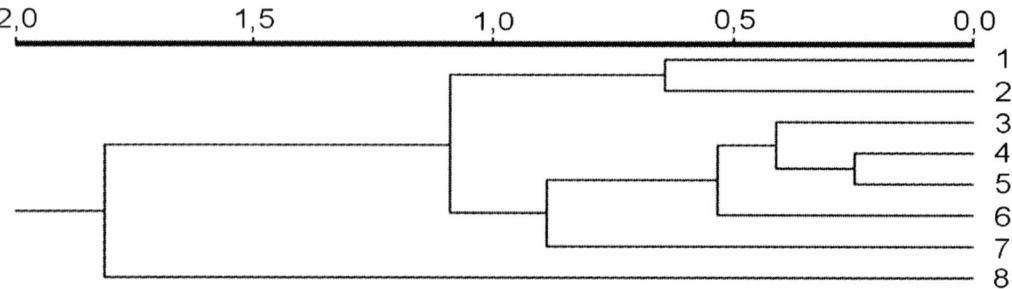

Kolyma River, 6 - lower reaches of Yenisei River), 7 – Lake Baikal (data by Yegorov, 1961) and sample of sterlet sturgeon from lower reaches of Yenisei River (8) based on morphometric characters.
Fig. 28. The consensus tree of the Siberian sturgeon samples (1 – middle reaches of the Lena River, 2 – Aldan River, 3 – delta of Lena River, 4 – lower reaches of Indigirka River, 5 - lower reaches of Kolyma

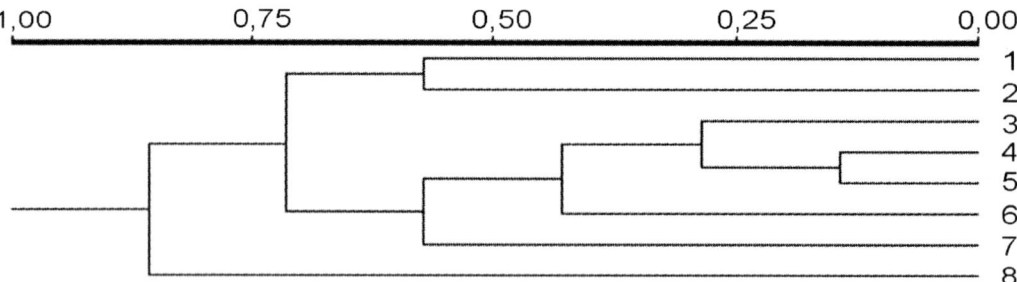

River, 6 - lower reaches of Yenisei River), 7 – Lake Baikal (data by Yegorov, 1961) and sample of sterlet sturgeon from lower reaches of Yenisei River (8) based on morphometric characters, constructed using single-link and complete-link dendrograms (fig. 26, 27).
Fig. 29. The single-link dendrogram of the Siberian sturgeon samples (1 – middle reaches of the Lena River, 2 – Aldan River, 3 – delta of Lena River, 4 – lower reaches of Indigirka River, 5 - lower reaches of Kolyma River, 6 - lower reaches of Yenisei River, 7 – lower Ob River) and sample of sterlet sturgeon from

with the opinion of other authors that the lower Lena is inhabited by a distinct and local population of sturgeon (Karantonis et al. 1956).

Clinal variability of meristic characteristics of Siberian sturgeon in the Ob and Yenisei basins is not as clearly expressed as it is in the Lena basin, because the data presented in Tables 7 and 8 were collected by various authors. However, some clinal variability of meristic traits similar to that described for the Lena basin is also observed. Sturgeon samples are characterized by large mean values of the number of scutes in the dorsal and lateral rows and the number of rays in the anal fin in the northern part of the Ob basin. Unfortunately, we were unable to evaluate the significance of the differences as not all authors reported a standard deviation. A similar trend is observed in the Yenisei basin: the mean number of scutes in the dorsal row and the number of rays in the dorsal and anal fins are larger in samples from the Yenisei delta and Yenisei Bay than in samples from the section of the river between Yartsevo and Vorogovo. However the significant differences are only in number of rays in the anal fin and the number of scutes in the lateral row while comparing the samples from the section of the river between Yartsevo and Vorogovo with samples from the river delta and the Yenisei Bay. In the first sample the mean number of dorsal scutes is significantly lower than in the sample from the Yenisei Bay. In contrast to the Ob and Lena sturgeon samples, the fish from the northern Yenisei (the Yenisei Gulf and the delta) are characterized by a lower number of scutes in the lateral row in comparison with sturgeon from the southern part (between Yartsevo and Vorogovo). These results are presented in Table 9. These data also indicate the existence of local, morphologically distinct sturgeon populations in the Yenisei and Ob basins.

The presented meristic characteristics of the Ob and Yenisei sturgeon are from the period prior to the dam construction at these rivers, when the spawning grounds extended from the mouth of the Chulym to the confluence of the Biya and Katun rivers (Ob). Including these last two rivers, the range of spawning grounds in the Ob basin extended approximately over 1,170 km (Dryagin 1949, Votinov 1963). Spawning grounds in the Yenisei River extend over an even greater range, from 400 km (Dryagin 1949) to 2,012 km upstream from the river mouth (at the city of Yeniseisk) and possibly up to the village Atamanovo, 2,256 km upstream (Podlesniy 1995) giving a total spawning range of 1,612 to 1,856 km. Such a significant extent of the spawning range, also confirm the hypothesis on the existence of morphologically distinct local populations of sturgeon in both rivers,

Clinal variation in meristic traits was not observed in the 700-km stretch of the lower Kolyma River (Tab. 12). According to our observations, the spawning grounds of this population extend only over a stretch of 250 km starting from Srednekolymsk (650 km upstream from the mouth) to the mouth of the Ozhogina River (900 km upstream), indicating that a single population inhabits the Kolyma. Probably an analogous situation exists in other rivers (the Yana and the Indigirka), where the sturgeon's range is not as extent and far upstream as in the major river systems (Ob, Yenisei, Lena).

The analysis of the meristic traits demonstrated (Tab. 11, 12) that there are differences in means of the traits between Siberian sturgeon populations inhabiting different river basins. Below only statistically significant differences will be analyzed. Samples from the lower reaches of the Ob have fewer dorsal scutes and gill rakers compared to those from the lower reaches of the Yenisei, Lena, Indigirka and Kolyma and a higher number of anal fin rays than those from the Aldan and the middle reaches of the Lena (see Tab. 11, 13).

Samples from the lower reaches of the Yenisei have the greatest number of dorsal fin rays and ventral and lateral scutes. The number of anal fin rays in sturgeon from the lower Yenisei is less than from the Lena delta and the lower Indigirka and greater than in the Aldan and the middle Lena samples. Similar comparative relationships between meristic characters such as dorsal, lateral and ventral scutes, as well as numbers of gill rakers, dorsal and anal fin rays, can be identified from the Tables 11 and 12 with the respective significance estimates in Table 13.

As was shown above, mean values of meristic characters in the study samples vary widely (Tab. 8, 9, 10, 11, and 12). With respect to meristic traits there are no limitations to one-dimensional comparisons as for morphometric characters. Thus it is possible to make paired comparisons between meristic traits belonging to the four previously described subspecies – *A. baerii baerii* from the Ob, *A. baeri stenorrhynchus* from the Yenisei, *A. baerii chatys* from Yakutian water and *A. baerii baicalensis* from the Lake Baikal basin – using Mayr's coefficient of difference (CD) (Mayr 1971). Values for CD from paired comparisons on meristics between our samples and literature data from upper reaches of the Ob (Petkevich *et al.* 1950) and Lake Baikal (Yegorov 1961) are presented in Table 14. From the given results, it is clear that between the samples belonging to the four previously described subspecies, the threshold of subspecies separation (1.28) is not achieved in Mayr's CD (Tab. 14) despite of large differences between mean values of some samples.

Cluster analysis using meristics to assess the degree of similarity of our samples from different parts of the species range by yielded dendrograms with the single-link method as shown in Figure 29 was performed. Therein samples from the lower reaches of the Indigirka, Kolyma, Lena and Yenisei rivers are combined into a single cluster within which the level of similarity corresponds to the distance between the rivers. The greatest similarity exists between samples from the Indigirka (4) and the nearby Kolyma (5). Less similarity exists in samples from the delta of the Lena (3) and the lower Yenisei (6). Samples from the southern sections of the range including the middle Lena (1) and it's tributary the Aldan (2) show a lower similarity with samples from the northern sections. The sample from the lower Ob (7) has the lowest similarity to others. As outgroup, the sample of sterlet from the Yenisei River was used (sample 8). Figure 30 shows a similar cluster in the complete-link dendrogram. In contrast from the previous dendrogram, samples from the southern parts of the Lena basin and the lower Ob join into separate cluster. As with the single-link dendrogram, the samples from the northern part of the habitat, with the exception of the Ob, comprise a single cluster.

The high correspondence of the single-link and complete-link dendrograms structure (Figs. 29

Table 14:
Values of coefficient of difference (CD by E. Mayr) between samples of the Siberian sturgeon, meristic characters.

Subspecies	A. baerii baerii	A. baerii stenor-rhynchys	A. baerii chatys					A. baerii baica-lensis	
Water body	Upper Ob (Petkevich et al. 1950)	Lower Yenisei (our data)	Lena delta (our data)	Aldan (our data)	Middle Lena (our data)	Lower Indigirka (our data)	Lower Kolyma (our data)	Lake Baikal (Yegorov 1961)	Characters
1	2	3	4	5	6	7	8	9	10
Lower Ob (our data)	0.23	0.35	0.18	0.05	0.19	0.18	0.21	0.35	D
	0.33	0.04	0.17	0.53	1.02	0.26	0.05	0.24	A
	0.26	0.42	0.48	0.13	0.18	0.47	0.53	0.51	Dr
	0.07	0.54	0.22	0.22	0.18	0.28	0.18	0.83	Lr
	0.51	0.45	0.19	0.14	0.08	0.13	0.09	0.92	Vr
		0.66	0.86	0.93	0.67	0.54	0.71	0.57	sp.br.
Upper Ob (Petkevich et al., 1950)		0.08	0.08	0.31	0.41	0.10	0.07	0.08	D
		0.36	0.50	0.26	0.83	0.63	0.38	0.56	A
		0.62	0.68	0.36	0.06	0.66	0.71	0.70	Dr
		0.67	0.31	0.31	0.27	0.39	0.27	1.00	Lr
		0.19	0.22	0.34	0.48	0.27	0.28	0.90	Vr
Lower Yenisei (our data)			0.19	0.46	0.55	0.21	0.18	0.01	D
			0.12	0.54	1.00	0.20	0.00	0.19	A
			0.04	0.27	0.55	0.05	0.10	0.10	Dr
			0.32	0.32	0.39	0.32	0.37	0.25	Lr
			0.27	0.34	0.42	0.30	0.30	0.48	Vr
			0.18	0.24	0.04	0.12	0.01	0.06	sp.br.
Lena delta (our data)				0.26	0.38	0.01	0.02	0.18	D
				0.66	1.10	0.07	0.12	0.07	A
				0.32	0.61	0.01	0.07	0.06	Dr
				0.00	0.06	0.03	0.05	0.59	Lr
				0.07	0.14	0.05	0.08	0.76	Vr
				0.06	0.23	0.30	0.18	0.23	sp.br.
Aldan (our data)					0.17	0.27	0.30	0.46	D
					0.54	0.79	0.56	0.71	A
					0.29	0.32	0.37	0.36	Dr
					0.06	0.03	0.05	0.58	Lr
					0.07	0.01	0.02	0.84	Vr
					0.29	0.36	0.24	0.28	sp.br.
Middle Lena (our data)						0.39	0.42	0.55	D
						1.25	1.03	1.14	A
						0.59	0.64	0.63	Dr
						0.09	0.01	0.68	Lr
						0.07	0.04	0.95	Vr
						0.09	0.05	0.03	sp.br.
Lower Indigirka (our data)							0.04	0.20	D
							0.21	0.01	A
							0.05	0.05	Dr
							0.08	0.62	Lr
							0.03	0.75	Vr
							0.13	0.06	sp.br.
Lower Kolyma (our data)								0.17	D
								0.19	A
								0.01	Dr
								0.65	Lr
								0.73	Vr
								0.07	sp.br.

and 30) and the consensus tree (Fig. 31) supports the robustness of the clustering structure.

Our material lacked samples from the Lake Baikal. However, the inclusion of data from Yegorov (1961) does not significantly alter the structure of the dendrograms. From Figure 32, in the single-link dendrogram, samples from the lower reaches of the Yakutian rivers are unified in a single cluster with a high level of similarity. Likewise, on a high level of similarity are united samples from the Yenisei basin, the lower reaches of the river and Lake Baikal. These two clusters are united at a lower level of similarity. Less similar yet are the samples from the southern part of Lena basin (Aldan and middle reaches of the Lena). Sturgeon sample from the lower Ob is connected at the lowest level of similarity. Thus the analysis of this dendrogram demonstrates that on one hand, the level of similarity of certain samples belonging to the previously distinguished subspecies *A. baerii stenorrhyncus*, *A. baerii chatys* and *A. baerii baicalensis* is relatively high. On the other hand, within the subspecies *A. baerii chatys* there are populations (inhabiting southern part of the subspecies range – Aldan and middle reaches of Lena), which level of similarity by meristic characters with populations from the northern part of the habitat range (Lena delta, lower Indigirka and Kolyma) is lower than between northern populations of different subspecies. The dendrogram plotted by the complete link method (Fig. 33) is of some difference but also shows a combination within of cluster of samples belonging to different subspecies – *A. baerii stenorrhynchus* from the Yenisei, and *A. baerii baicalensis* from the basin of Lake Baical – and, at a slightly lower level of similarity, *A. baerii chatys* from the Indigirka, Kolyma and the lower Lena. The samples of sturgeon of the latter subspecies from the southern part of the Lena basin (Aldan and middle reaches of Lena) are combined with the sample from the lower reaches of the Ob (*A. baerii baerii*). It should be noted that the level of similarity between samples from the Aldan and the middle Lena is somewhat lower than between *A. baerii stenorrhynchus* and *A. baerii baicalensis*.

The consensus tree based on the single-link and complete-link trees (Fig. 34) has a similar structure and demonstrates their high similarity.

The results of the cluster analysis of the samples of Siberian sturgeon by meristic and morphometric characters lead to the conclusion that identical results were obtained for both sets of criteria. The similarity between samples from the northern part of the habitat of the subspecies *A. baerii baerii*, *A. baerii stenorrhynchus*, *A. baerii chatys* and also *A. baerii baicalensis* is relatively high. Furthermore, the similarity between samples from the northern part of the habitat of one subspecies (*A. baerii chatys* from the lower reaches of the Lena, Indigirka and Kolyma) and samples from this same subspecies from the southern part of the Lena basin, the Aldan and middle reaches of the Lena, was lower than between samples of different subspecies from the northern part of the habitat.

3. The ecological forms of the Siberian sturgeon

It is well known that Siberian sturgeons besides riverine forms are also represented by lacustrine-riverine forms (Lake Baikal, Lake Zaisan before damming at Irtysh river). Besides that there has been a long lasting debate in the literature on resident, anadromous or semi-anadromous forms. While the existence of resident forms (races) has never been doubted, the degree of anadromy is debated. The relatively small basins of the Yana, Indigirka, and Kolyma rivers are inhabited by typical riverine populations (Sokolov and Kashin 1965). According to our observations, the number of sturgeons in the deltas of these rivers is rather low. The fish primarily resides in areas higher upstream. These observations agree with the results of other authors (Kirillov 1972, Rybnikov 1961).

With regard to the migrations of the Lena River sturgeon we have contradictory data. Several authors associate the species with typical riverine forms (Pirozhnikov 1955, Sokolov 1966b and 1981, Sokolov and Kashin 1965) not undergoing long-distance migrations but remaining more or less "settled" (Sokolov 1981). The lengths of these movements have been determined by tagging and are in the order of only tens or hundreds of kilometres (Dormitontov and Safronov 1976). According to P. L. Pirozhnikov's (1955), the Lena population, with the exception of a small number of individuals, spend its entire life in the river. Gonad maturation occurs in places of its more or less constant habitat, particularly in places of wintering. There are data that besides the well-known spawning grounds in the lower reaches of the Lena River (which are not too far from the mouth of the Natara), there are also spawning grounds of the sturgeon located further downstream between the village of Chekurovka and Stolb Island and in the upper reaches of the Trofomovskaya and Bykovskaya branches of the Lena delta (Pirozhnikov 1955, Dormitontov 1963). Pirozhnikov considers the summer-fall upstream migration in the lower reaches of the Lena different from an analogous migration among anadromous and semi-anadromous fish because in the former case all individuals migrate whereas in the latter only those fish migrate who are mature and ready to spawn. It should be noted that the spawning grounds and the feeding grounds of the lower Lena are contiguous. Outside the spawning season sturgeon also feed in the spawning grounds, though less intensely than in the branches of the delta. Other authors claim that a semi-anadromous form of sturgeon exists in the Lena, feeding in the estuary and in the delta and migrating up to 1,650 km upstream (Sokolov and Vasiliev 1989). In order to designate this population, Dryagin uses the neutral term "migratory form" (Dryagin 1949). He notes that the range of this form overlaps with the habitat of the resident population probably between the city of Yakutsk and the mouth of the Viluy River with prevalence of the resident form upstream the Yakutsk. However, he does not support these conclusions with any data. The term "migratory form" seems inadequate in the given case, as it can be applied to almost all fish populations because spawning, feeding and wintering grounds almost always do not completely or partially coincide and fish have to perform regular movements i.e. migrations.

According to Dryagin (1949), resident and

migratory forms of sturgeons exist in the Yenisei system as well as in other large river systems (Ob, Irtysh, Lena). The first is a riverine form that always resides in the middle and upper sections of the rivers and does not migrate onto shoals and deltas. Immature specimens of the second, migratory form, migrate downstream to shoals and deltas to forage and winter and re-enter the river to spawn. Dryagin particularly notes that firstly, the migratory form stays in fresh water its entire life and only occasionally can individuals be found in brackish waters (5 to 10 ppm), and, secondly, the migrations of this form are not as distinct as in anadromous species where all juveniles migrate at roughly the same age and where a distinct spawning migration of all fish can be identified.

The Yenisei sturgeon form named by Dryagin "migratory" is considered by other authors to be "semi-anadromous" (Sokolov and Kashin 1965). Having never accepted the "semi-anadromous" concept, A. V. Podlesniy (1954, 1955, 1958, 1963, 1968) was concerned that an anadromous form of the sturgeon exists in the Yenisei River. Subsequently other authors adopted this opinion (Barannikova 1975, Yegorov 1988, Mikhalyov 1967 and 1969, Mikhalyov et. al 1975). However, if one agrees with Podlesniy's opinion on the existence of both resident and anadromous forms in Yenisei, than this is confusing. First, the sturgeon does not enter salt water, so in the best case it can only be said to be semi-anadromous. It should be noted that the sterlet, which the author does not consider to be anadromous, also migrates to the southern end of the Yenisei Gulf (Berg 1949). Second, the range of the resident and so-called "anadromous" forms overlap, according to different accounts, over an extensive portion of the river (1,000-1,400 km) (Podlesniy's 1955). This, with assertion on the morphological indistinguishability of these forms (Podlesniy's 1955 and 1958, Mikhalov 1967), makes the assumption rather absurd. Third, data on the spawning range of the sturgeon are presented without an indication of the spawning grounds of separate forms.

Given these facts there is no reason to speak on the extended migration of a so-called "anadromous" form of Siberian sturgeon overlapping in range with resident forms. According to combined data of various investigators spawning grounds in the Yenisei are distributed over a particularly large section of the river about 1,612 to 1,856 km in length (see above). This, together with the data presented in Chapter 3 on the clinal variation of several meristic traits, supports the hypothesis of geographically separated local populations of sturgeon in the Yenisei as was demonstrated in the Lena. Apparently, in the lower reaches of the Yenisei, a separate population of sturgeon exists that forages in the Yenisei Bay and the Gulf of Yenisei as well as in the lower reaches of the river and that shares the lower section of the spawning range.

Identifying any form to belong to anadromous, semi-anadromous or resident fishes depends on accepted definitions and classifications. Podlesniy (1968), defined a species as anadromous by the ability to osmoregulatory, allowing to switch from freshwater to sea water, thereby distinguishing two groups of anadromous fish in the Yenisei. To the first he appropriated the omul (*Coreganus autumnalis*), vendace (*C. albula*), muksun (*C. muksun*), and smelt (*Osmerus* sp.), all of which forage in the Gulf of Yenisei and the Yenisei Bay and display a distinct spawning migration without exhibiting resident freshwater forms. To the

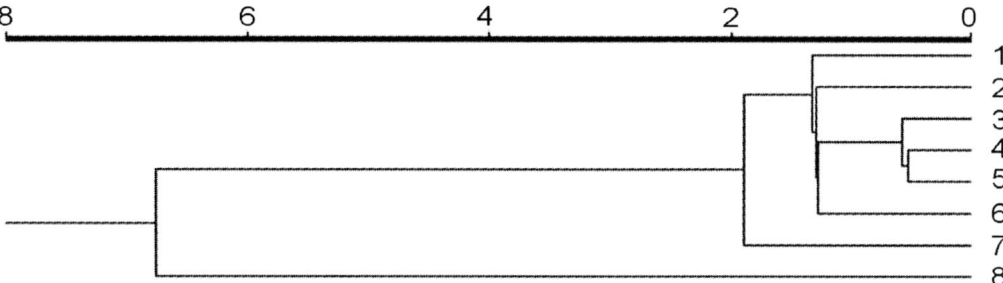

lower reaches of Yenisei River (8) based on meristic characters.
Fig. 30. The complete-link dendrogram of the Siberian sturgeon samples (1 – middle reaches of the Lena River, 2 – Aldan River, 3 – delta of Lena River, 4 – lower reaches of Indigirka River, 5 - lower reaches of Kolyma River, 6 - lower reaches of Yenisei River, 7 – lower Ob River) and sample of sterlet sturgeon from

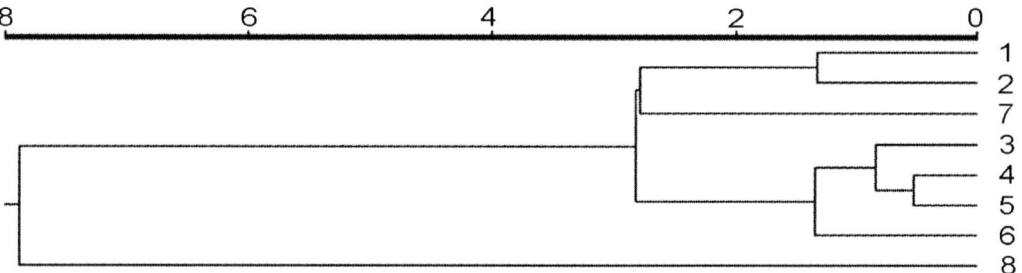

lower reaches of Yenisei River (8) based on meristic characters.
Fig. 31. The consensus tree of the Siberian sturgeon samples (1 – middle reaches of the Lena River, 2 – Aldan River, 3 – delta of Lena River, 4 – lower reaches of Indigirka River, 5 - lower reaches of Kolyma River, 6 - lower reaches of Yenisei River, 7 – lower Ob River) and sample of sterlet sturgeon from lower

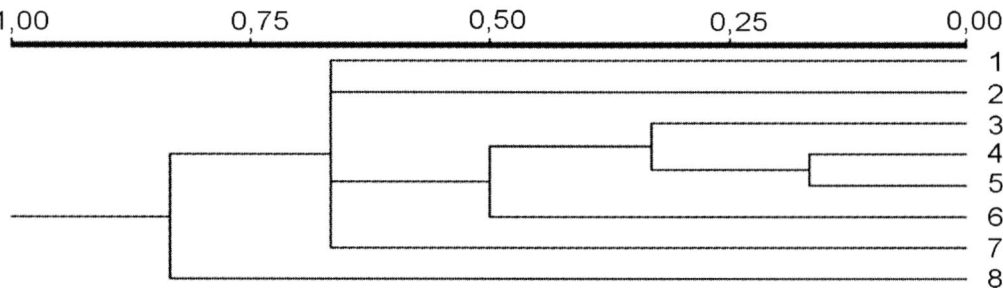

reaches of Yenisei River (8) based on meristic characters, constructed using single-link and complete-link dendrograms (fig. 29, 30).
Fig. 32. The single-link dendrogram of the Siberian sturgeon samples (1 – middle reaches of the Lena River, 2 – Aldan River, 3 – delta of Lena River, 4 – lower reaches of Indigirka River, 5 - lower reaches of Kolyma River, 6 - lower reaches of Yenisei River, 7 – lower Ob River, 8 - Lake Baikal (Yegorov, 1961)) and

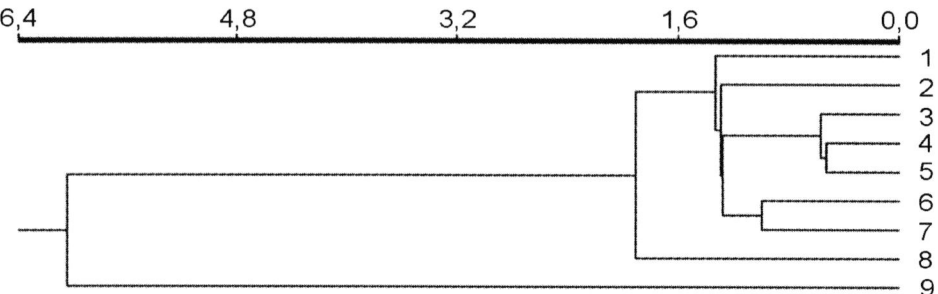

sample of sterlet sturgeon from lower reaches of Yenisei River (9) based on meristic characters.
Fig. 33. The complete-link dendrogram of the Siberian sturgeon samples (1 – middle reaches of the Lena River, 2 – Aldan River, 3 – delta of Lena River, 4 – lower reaches of Indigirka River, 5 - lower reaches of Kolyma River, 6 - lower reaches of Yenisei River, 7 – lower Ob River, 8 - Lake Baikal (Yegorov, 1961)) and

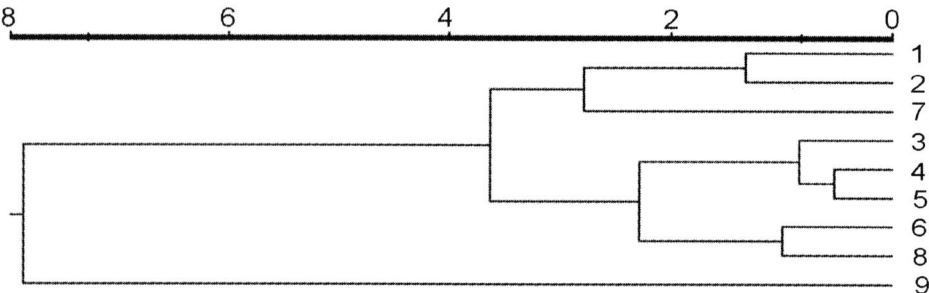

sample of sterlet sturgeon from lower reaches of Yenisei River (9) based on meristic characters.
Fig. 34. The consensus tree of the Siberian sturgeon samples (1 – middle reaches of the Lena River, 2 – Aldan River, 3 – delta of Lena River, 4 – lower reaches of Indigirka River, 5 - lower reaches of Kolyma River, 6 - lower reaches of Yenisei River, 7 – lower Ob River, 8 - Lake Baikal (Yegorov, 1961)) and sample

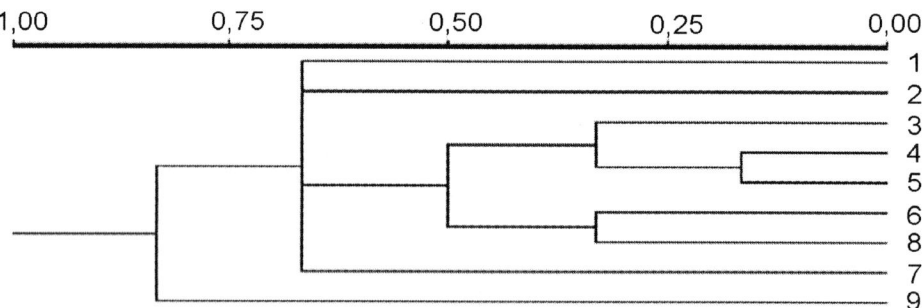

of sterlet sturgeon from lower reaches of Yenisei River (9) based on meristic characters, constructed using single-link and complete-link dendrograms (fig. 32, 33).

second group, exhibiting both anadromous and resident forms, Podlesniy attributed sturgeon, white salmon (*Stenodus leucichthys nelma*), and whitefish (*C. lavaretus*), whose spawning migration is not as distinct as in the first group. The juvenile downstream migration of the second group lasts several years. Certain individuals of "anadromous" sturgeon don't undertake the migration until their twentieth year (Podlesniy 1954) or just before reaching sexual maturity, it is essentially impossible to distinguish this form from the resident form.

Most distinct migrations of Siberian sturgeon are displayed in the Ob-Irtysh basin were (besides resident populations) the so-called anadromous form occur (Votinov 1963, Votinov et. al. 1975, Dryagin 1949, and others). This migratory form was considered by other authors as semi-anadromous (Petkevich and Ioganzen 1958). The existence of this form is, in fact, compelled by a unique geophysical feature of this basin, namely, the annual winter hypoxia, which forces all fish from the lower and middle reaches of the river to migrate into the Ob bay in the beginning of winter. These fish then return to the river to forage and reproduce in the spring, after the ice melts and oxygen levels improve to permit survival. In contrast to true anadromous species of sturgeons that forage in saltwater and migrate into rivers only to spawn, the migration of the Siberian sturgeon from this particular bay into the river is a feeding migration because all age and size groups migrate. Only for fish preparing to spawn does this migration later become a spawning migration, which can last over a year. When the upstream migration along the Ob and Irtysh is completed, they remain above the hypoxic zone to overwinter (Votinov 1963). Before the construction of the dam, the extent of their migration was rather great – in the Ob, migrations extended over 3000 km to the confluence of the Biya and Katun Rivers where the Ob begins; in the Irtysh they extended to the mouth of the Bukhtarma (Sokolov and Vasiliev 1989). Currently the spawning migration of this form is limited by the dams of the Novosibirsk Hydroelectric Station on the Ob and the Shulbinskaya Hydroelectric Station on the Irtysh (Votinov 1962, Votinov et al. 1975, *The Fish of Kazakhstan* 1986). Along with the "anadromous" form of sturgeon, the Ob basin is inhabited by a resident form, the range of which presumable overlapped with the range of the "anadromous" form over the entire length of the Ob River upstream of the mouth of the Irtysh and most of the Irtysh River (Dryagin 1949). In the Irtysh basin above Lake Zaisan there was a supposed resident Zaisan-Chernyi-Irtysh population (Bogan 1938, Dryagin 1948a and b 1949), though a distinct boundary between the "anadromous" and resident forms was not established (*Fish of Kazakhstan* 1946). After hydroelectric dams were constructed in the Irtysh, the Zaisan-Chernyi-Irtysh population was separated and each of the reservoirs is now inhabited by an isolated population – the Bukhtarminskaya and Ust-Kamennogorskoaya populations (*Fish of Kazakhstan* 1986). Presumably, an analogous population inhabits the section of the river between the Shulbinskaya and Ust-Kamennogorskaya Hydroelectric Stations.

Thus, attributing of the Siberian sturgeon forms, inhabiting the lower reaches of large Siberian rivers (Lena, Ob, Yenisey), feeding in brackish waters of bays and deltas of these rivers, to anadromous, semi-anadromous or resident freshwater form depends on accepted classification schemes and this can be illustrated with a series of examples.

According to the definition given by K. F. Kessler (1877, p. 313), a Siberian sturgeon is a semi-anadromous fish: "... some which constantly reside in rivers and freshwater lakes, while some reside in semi-saline lakes or in river mouths, but in last case migrate to rivers to spawn, though without undertaking migrations as extensive as those of true anadromous fish." Ioganzen (1947) classifies the Ob sturgeon as "anadromous" fish (they inhabit the Ob Bay and adjacent parts of the Kara Sea, from which they enter the river system in order to spawn). Ioganzen (1947) considered the sturgeon formerly inhabiting Lake Zaisan to be a "general-freshwater" fish because it always resides in freshwater lakes or rivers. According to V. I. Vladimirov (1957) the Siberian sturgeon from the lower reaches of the Lena, Yenisei and Ob should be considered a "semi-anadromous indigenously freshwater" fish. According to Burmakin and Tyurin (1959), these populations of sturgeon as well as those of the sterlet inhabiting the lower reaches and the deltas of the river should be considered "anadromous brackish-water" fish, other populations of Siberian sturgeon and sterlet are considered non-anadromous. The Baikal sturgeon must be considered a freshwater anadromous fish. According to Meissner's classification (1933), used later by Schmidt (1936), populations of Siberian sturgeon inhabiting the lower reaches of rivers and entering brackish bays to forage must be attributed as "generatively rheophilic, semi-anadromous, delta fish" whereas resident populations in the rivers are considered "generatively rheophilic freshwater fish of flowing waters". The Baikal sturgeon must be classified as a "generatively rheophilic freshwater fish of stagnant waters". The term "generative rheophilic" here means that fishes spend they spawning period in flowing waters of rivers. According to Podlesniy (1954, 1955, 1958, 1963, 1968) the Yenisei sturgeon is anadromous. More information on sturgeon migrations can be found in Pavlov *et al.*, 2001.

This diversity of attributions reflects the evolution of conception of classifying migrations and migratory cycles and also is an attempt to systematize the entire diversity of fish forms basing on their migratory cycles. At the same time the examples above show not only the complexity of the problem but also the imprecision of the criteria for distinguishing between separate biological groups, as a result of which the same population (e.g. the Baikal sturgeon) can be considered a "generatively rheophilic freshwater fish of stagnant waters" or "freshwater anadromous". It seems reasonable to accept the opinion of D.A. Shubnikov (1976) who, in accordance with Podlesniy (1968), identifies the existence of the osmoregulatory apparatus, which allows fishes to change freshwater environment on marine one and vice versa, as the only reliable trait to characterize "anadromous" fish. Thus, one can not consider a freshwater fish, making the migration from a lake to a river (such as in the case of the Lake Baikal Siberian sturgeon), to be "anadromous" even if it performs extensive migrations. Likewise, the classification of the Siberian sturgeon populations that reside in the lower reaches of the major rivers and forage in brackish bays as "anadromous" seems insufficiently justified, since, as stated above, the sturgeon never leave the limits of significantly freshened waters. This is further supported by experimental data showing that upon transfer into a hypertonic medium, Siberian sturgeon are incapable to adapt their type of osmoregulation to a hypotonic regime while their ability to withstand saltwater is comparable to that of other freshwater fish (Krayushkina and Moiseyenko 1977).

4. Conclusions

Morphological traits among Siberian sturgeons inhabiting an exceptionally broad geographic range within a wide range of environmental conditions are highly variable. Morphometric characters from different watersheds attest to the complexity of the species' phenotypic structure. It's notable peculiarity is the relatively high level of similarity of populations living in the lower reaches of the rivers across its entire geographical range. There is no separation when using principal component analysis for samples belonging to formerly described subspecies *A. baerii baerii* and *A. baerii stenorrhynchus* and *A. baerii chatys* all collected along the northern margin of the distributional ranges, i.e. the lower reaches of the Ob, Yenisei, Lena, Indigirka and Kolyma basins. Within the Lena basin, according to morphometric characters were distinguished phenones, the level of difference between which is greater than between the geographically isolated populations of formerly separated subspecies. Sturgeon samples previously attributed to the subspecies *A. baerii chatys* from the southern end of the subspecies range (the Aldan and middle Lena) and from the northern end (the lower Lena, Indigirka and Kolyma), are more distinctly separated by principal component analysis than the northern samples of *A. baerii baerii* (lower Ob), *A. baerii stenorrhynchus* (lower Yenisei), and *A. baerii chatys* (lower Lena, Indigirka and Kolyma).

Cluster analysis of Siberian sturgeon samples using meristic and morphometric characters yield nearly identical results. As with the Principal Component analysis, the level of similarity among northern samples of *A. baerii chatys* from the lower Lena, Indigirka and Kolyma, and samples of sturgeon of this same subspecies from the southern ends of the distributional range (Aldan and middle Lena) is lower than between northern samples of sturgeon from subspecies *A. baerii baerii* and *A. baerii stenorrhynchus* and the Baikal sturgeon *A. baerii baicalensis*. These results together with paired comparison tests according to meristic traits using Mayr's coefficient of difference attest to the invalidity of the separation of these subspecies in Siberian sturgeon.

The results of the analysis of phenotypic diversity of the populations of Siberian sturgeon according to morphometric and meristic traits suggest that this species is monotypic.

Within the Lena basin, Siberian sturgeons demonstrate a directional clinal variability in meristic and several morphometric characters from north to south, coinciding with changes that occur when northern Lena sturgeons are cultivated in warm-water aquaculture. This attests to the temperature dependence of the observed clines. Furthermore, time and places of spawning in the Lena basin infer the existence of local morphologically differentiable populations. Analogous clines for the same traits exist in other major river basins (e.g. Yenisei and Ob).

Thus, the monotypic species *Acipenser baerii*

forms independent populations inhabiting separate rivers and river systems: Ob, Yenisei, Katanga, Pyasina, Anabar, Olenyok, Lena, Yana, Indigirka, Kolyma. However, the existence of a continuous row of populations or a "population continuum" within the major river systems (Ob, Yenisei, Lena) can be considered as ranked populations (Mina 1986). Just in population continuums clinal variations as a rule were observed (Mayr 1971).

An analysis of data on different ecological forms of Siberian sturgeon – riverine, lacustro-riverine, resident, and those that undergo extensive migrations – the location and time of their spawning, their capacity to adapt to various salinities, and data on clinal variability of morphological traits, allows to conclude that the Siberian sturgeon cannot be considered as anadromous or even semi-anadromous. This is in contradiction to the opinion of many authors (Podlesniy 1954, 1955, 1958, 1963, 1968, Sokolov and Kashin 1965, Mikhalyov 1967 and 1969, Mikhalyov et al. 1975, Barannikova 1975, Yegorov 1988, Sokolov and Vasiliev 1989). The entire life cycle of the Siberian sturgeon is bound to freshwater; its populations inhabiting the lower reaches of rivers do not leave the limits of fresh or weakly brackish waters. Thus, the migrations of Siberian sturgeon, including the Ob and Baikal populations, whose migrations are extensive, can be considered potamodromous.

The principal differences between the annual migrations of the Ob sturgeon from the river into the Ob Bay in the fall and early winter and back upstream in the spring, and the migrations of truly anadromous species that feed in saline marine waters and migrate to rivers only for spawning, can be interpreted as follows. First, these annual migrations, which are undertaken by the Siberian sturgeon as well as several other freshwater fish including the sterlet, *Leuciscus idus* and others, are brought about by the annual winter hypoxia event that forces fish to move into the Ob bay or to move upstream to remain above the hypoxic front. Second, the Siberian sturgeon in the Ob bay remains within fresh or weakly saline waters. Finally, in contrast to the truly anadromous acipenserids, among which only fish preparing to spawn migrate, this migration of the Ob sturgeon includes fish of all ages and size classes and in reality is a feeding migration. Only for those fish preparing to spawn does this migration later become a spawning migration (Votinov 1963).

In summary, the Siberian sturgeon is a monotypic species represented by isolated populations inhabiting separate river systems (Ob, Yenisei, Khatanga, Pyasina, Anabar, Olenyok, Lena, Yana, Indigirka, Kolyma). In the major river basins (Lena, Yenisei, and Ob) this species exhibits a continuous row of populations (population continuum). Within its range, the Siberian sturgeon has riverine and lacustro-riverine forms, some of which are resident within a local area while others make extensive potamodromous migrations. The structure of the phenotypic diversity of the Siberian sturgeon is determined by parallel latitudinal clinal variability in morphometric and meristic traits among the populations of the major river basins (i.e. the Lena, Yenisei, and Ob).

The explanation for the existence of monotypic freshwater representatives of the genus *Acipenser* in the territory of Siberia – the Siberian sturgeon inhabiting virtually every major river basin - is necessarily to be found in sturgeon phylogeny.

The paleontological history of acipenseriforms remains insufficiently studied. There is no single opinion on time and place of their origin. Some authors (Miller 1969) believe that the *Polyodon*, *Scaphyrinchus* and *Acipenser genus* originated at the territory of modern North America corresponding in present time to the Mississippi basin. Others (Yakovlev 1977) believe that freshwater *Chondrostidae*, distributed at the beginning of the Jurassic era across the entire Palearctic region, attaining moderate size and an archaic structure, gradually evolving in the direction towards modern *Acipenseriformes*.

According to the most recent data, acipenseriforms emerged in the Jurassic epoch in the basin of the ancient Thetis Sea in modern Central Asia (Birstein and DeSalle 1998). The emergence of acipenserids as a result of the divergence of *Acipenseridae* and *Polyodontidae* is presumably due to a divergence from an extinct ancestor (*Peipiaostidae*) known from late-Jurassic sedimentary structures (Jin 1995, Jin et al. 1995, Grande, Bemis 1996) and also associated with the Jurassic period 200-135 million years ago (Birstein and DeSalle 1998).

The most ancient fossil records of acipenserids data from the late Cretaceous, 95-65 million years (Nesov and Kaznyshkin 1983).

Currently, there is no consensus regarding the phylogenetic connections within the Acipenseridae family (Berg 1904 and 1905, Findeis 1993 and 1997, Mayden and Kuhajda 1996). The phylogenetic connections within the *Acipenser* genus have also not yet been clarified, although over a long time there were attempts to group species together (see the reviews of Dumeril 1870 and Bemis et al. 1997). Such attempts, based on biogeographical data (Artyukhin 1995) and molecular genetics (Birstein et al. 1997, Birstein and DeSalle 1998) were undertaken during the past few years.

Molecular-genetic investigations (Berstein and DeSalle 1998) indicate that the closest relative to the Siberian sturgeon *Acipenser baerii* is the Persian sturgeon *A. persicus* and the Adriatic sturgeon *A. naccarii*. Less genetic similarity has been reported between the Siberian sturgeon and the Russian sturgeon (*A. gueldenstaedtii*). Between the last two species the morphological similarity was noted (Sokolov and Vasiliev 1989).

Following the opinion expressed by Birstein and DeSalle (1998) the ancestor of the Siberian sturgeon entered the Siberian rivers from the Ponto-Caspian basin during the middle of the Pleistocene through the system of peri-glacial lakes that existed during maximum glaciations (Berg 1928). The eastern and south-eastern limits of the modern range of the Siberian sturgeon are determined by the extent of the last glaciation that occurred 18,000-20,000 years ago (Velichko 1994) and from the late Pleistocene glaciations of 7,000 years ago to this day have not changed (Tzepkin 1995).

Thus, the monotypicity of the Siberian sturgeon is presumably a result of this being a phylogenetically young species of acipenserids, and it's freshwater lifestyle is a result of the evolution of the species during isolation from marine basins during the Ice Ages.

Part II: Ecological characteristics of the Siberian sturgeon

1. Geographical distribution of the Siberian sturgeon and it's changes under anthropogenic influences

The range of the Siberian sturgeon is exceptionally broad, from latitudes of 73-74°N at the mouth of the Lena and Ob Bay to 48-49°N in the Chernyi Irtysh, and Selenga rivers (Dryagin 1948a, Votinov et al. 1975) and longitudinally over 97° (Dryagin 1948a). The range of the Siberian sturgeon and its changes during the last 150 years are presented in Figure 35.

In the **Ob River basin** the northern boundary of distribution is at the Ob Bay near Cape Drovyanoy (Yudanov 1935, Dryagin 1948b and 1949). In the Ob itself, the species is encountered along its entire length, (3,680 km from the confluence of the Biya and Katun rivers (where the Ob originates) to its delta. In the Katun River it inhabits the area from the mouth to 50-70 km upstreams were spawning grounds have been identified. In the Biya River it is rare and only near the river mouth (Dryagin 1949, Petkevich et al. 1950). Specimens have also been caught in Lake Teletskoe (Berg 1948). Currently, there is some reduction of the range in the upper Ob and its tributaries. It is no longer found in the Biya river basin and Aley river and also not in the Katun upstreams Srostki village (53 km from the mouth). Habitat reduction here is due to pollution of the Biya (Biysk city), gravel extraction at the near-mouth of the Katun, pollution from sewage runoff from Rubtsovsk into the Aley and illegal fishing (Solovov 1997).

The Siberian sturgeon inhabits tributaries of the Ob River including the Chulym, Charysh, Nadym and Irtysh rivers, sometimes it can be found in the mouths of other large tributaries, the Polui and Synya rivers (Dryagin 1948b 1949). In the Chulym River, the sturgeon ranged up to 1,300 km upstream from the mouth (Khokhlova 1953). Siberian sturgeon range extends the entire length of the Irtysh River, from the mouth through Lake Zaisan and on up the Chernyi Irtysh River into the territory of China to the confluence with the Kren River (Bogan 1938, Dryagin 1948b, 1949, Petkevich et al. 1950, Sedelnikov 1910, Votinov 1963, Votinov et al. 1975). The sturgeon also inhabits an Irtysh tributary, the Tobol River, and the Tobol's tributaries, the Tura and Tavda rivers (Dryagin 1948b, 1949).

Taz River basin. Historically, rather large quantities of sturgeons were caught in the upper reaches of the Taz River, the caviar of which was transported down the Yenisei (Skalon, 1931). However, by the time of Skalon's work (1929-1930), sturgeon in the Taz became already rare. Votinov (1963 p. 7) wrote: "... if there was formerly a distinct Taz population of Siberian sturgeon, then by the 1930's it was already exterminated by fishing. The contemporary Taz river stock must be considered as a part of the Ob population and Taz Bay and lower reaches of rivers flowing into it as growing grounds for young sturgeon and foraging grounds for adults." In a later publication, Votinov et al. (1975) reported that the Siberian sturgeon does no longer occur in the Taz. However, in 1979 Siberian sturgeon were still reported to inhabit the Taz up to 300 km from the river mouth (Chupretov

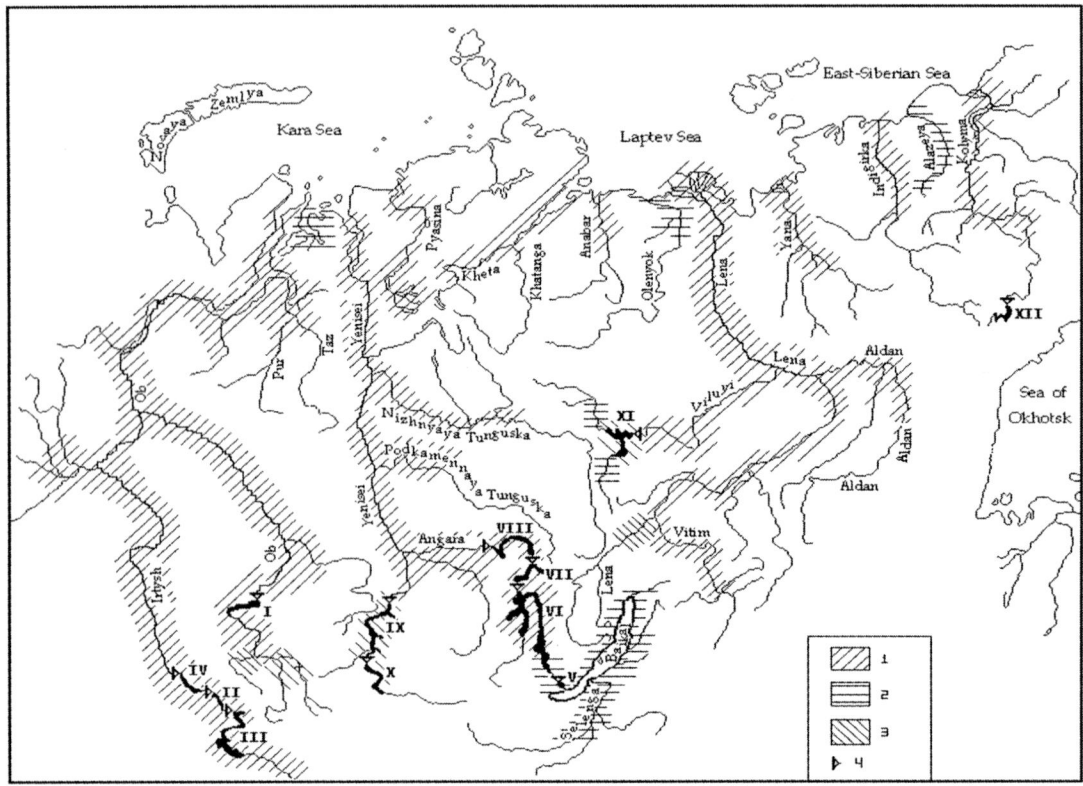

Fig. 35: Distributional range of the Siberian sturgeon.
Designations:
1 – Species usually occurs (common);
2 – Species occurs rarely (rare);
3 – In recent years does not occur, but formerly occurred (extinct);
4 – Dam of hydroelectric power station.
Reservoirs:
I – Novosibirskoye (constructed in 1957-1959);
II – Ust-Kamenogorskoye (constructed in 1950-1954);
III – Bukhtarminskoye (constructed in 1960-1964);
IV – Shulbinskoye (constructed in 1987);
V – Irkutskoye (constructed in 1956);
VI – Bratskoye (constructed in 1966);
VII – Ust-Ilimskoye (constructed in 1974-1977);
VIII – Bogutchanskoye (constructed in 1978);
IX – Krasnoyarskoye (constructed in 1967-1970);
X – Sayano-Shushenskoye (constructed in 1978);
XI – Viluyiskoye (constructed in 1965-1976);
XII – Kolymskoye (constructed in 1981-1984).

and Slepokurov 1979), while also occurring throughout the Taz Bay and occasionally in the near-mouths parts of Taz Bay tributaries: the Messo-Yakha, Anti-Payuta and Ader-Payuta rivers. The Siberian sturgeon also inhabits the lower reaches of the Pur River up to 100 km from the river mouth.

In Gyda Bay and the Gyda and Yuribey rivers the Siberian sturgeon is rare (Burmakin 1941, Dryagin 1949).

Yenisei River basin. The northern limit of the Siberian sturgeon range is Shirokaya Bay in the Yenisei Gulf. Before hydro- dam construction, the southern limit was the village of Oznachennoe (now Sayanogorsk city). This range covers a distance of 3,130 km to 3140 km (Podlesnyi 1955, Podlesnyi 1958, Dryagin 1949) or 3,215 km (Podlesnyi 1963). Earlier, Dmitriev (1941) considered the city of Minusinsk, located downstreams the Oznachennoe village to be the southern limit.

In the major tributaries of the Yenisei River – the Angara, Podkamennaya Tunguska and Nizhnyaya (Lower) Tungska rivers there are small resident populations of the Siberian sturgeon (Podlesnyi 1955, 1958). In the latter two rivers, the upper limits of the range are not indicated. Earlier, it was believed that the Yenisei River is inhabited by two forms of Siberian sturgeon, a resident and a so-called anadromous form. According to Podlesnyi (1955) opinion the resident form inhabited the Yenisei from the village of Oznachennoe to the city of Igarka (2,300 km). With respect to the "anadromous" form it was pointed out that it probably doesn't migrate, by one data above the village of Yartsevo (1,759 km from the river mouth) (Podlesnyi 1955) and by other data it moves above the town of Yeniseisk (Podlesnyi 1963). Upstream (2,454 km) from the city Krasnoyarsk, sturgeon were very rare (Podlesnyi 1963). Sturgeon also migrated to sites near the mouths of the Tuba and Abakan rivers (middle Yenisei tributaries) while they have also been observed in the lower reaches of the tributaries of the Nizhnyaya (Lower) Tunguska: the rivers Kochechumo, Vivi, Kuchumdek and Tutokchan (Dryagin 1949). In the XIX century, they were recorded in Kureyka River as well (Tretyakov 1869).

In the Angara basin, the distribution of the sturgeon differed before and after construction of Irkutskaya, Bratskaya and Ust-Ilimskaya hydrodams. Before the dams were built, Dryagin (1949), relying on observations by N. N. Bobrova, pointed out that in the Angara, the sturgeons were mainly found from the mouth to the confluence with the Oka River, and that individuals would occasionally be encountered up to the mouth of the Belaya River, 140 km from the source of the Angara. Furthermore, there have been isolated cases of young sturgeon being caught at the source of the Angara near Irkutsk, where they might have been drawn in by the rapid flow from Lake Baikal (Yegorov 1941, 1961). Sturgeon were also encountered in the Angara river tributaries the Taseeva river (and also in its tributary Chuna River) and Oka River. In the last river the sturgeon spawning site was presumably located (Yegorov, 1963). Thus, in spite of the fact that before dam construction at the Angara River the isolation of Baikal sturgeon and Yenisei sturgeon was significant, there was no reason to concern the range of Siberian sturgeon to be interrupted in the Yenisei river basin (including the Baikal Lake system). In Bratsk Reservoir, the sturgeon is distributed everywhere up to the upper reaches of the Oka and Angara rivers and enters the Belaya and Irkut rivers. In the Oka, the sturgeon is concentrated in the middle reaches where wintering holes are located (Lukyanchikov 1967b).

In Lake Baikal, Siberian sturgeon habitats are mainly located near the mouth of the mean tributaries. They are most abundant near the delta of the Selenga River and so called Selenga shoal as well as in the Barguzinskii and Chivyrkuiskii bays. They are less abundant in the northern part of Lake Baikal at the mouth of the Verkhnaya (Upper) Angara and Kichera rivers. The sturgeons move from these main habitats along the shallow littoral zones of the lake.

The Baikal sturgeon mainly enters the major rivers of the lake basin where spawning sites are located. Along the Selenga it migrates up to 1,000 km, entering such tributaries as Chikoi and Orkhon (Yegorov 1961) and the Tula and Delger-Muren rivers (Sokolov and Shatunovskii 1983). Presumably, a resident form exists in the Selenga and Orkhon rivers (Dashi-Dorzhi 1955). In the Bagruzin, the sturgeon travels over 300 km, and in the Verkhnaya Angara up to 100-150 km. There are data that sturgeons occasionally enter smaller tributaries, such as the Turka river (Yegorov 1961).

Pyasina River basin. Sturgeon was of low abundance in the Pyasina River (Dryagin 1949) although all age classes have been encountered. They have not been found in Pyasina Bay (Ostroumov 1937), though they are known to have inhabited lakes of the Pyasina basin, e.g. Lake Lama and Lake Melkoye (Belykh 1940, Logashov 1940). In recent years, sturgeons have almost disappeared from the Pyasina River and are only very rarely found in the Lama River (Savvaitova 1994).

Khatanga River basin. Earlier it was believed that Siberian sturgeon inhabited only the estuary of the Khatanga River (Berg 1926, Tetriakov 1869). Mikhin (1941) recorded sturgeon in the Khatanga, Kheta and presumably Kotui. There are controversial opinions on the place of Khatanga River forming. Following L. K. Davydov (1955) Khatanga is formed at the confluence of the Moyero and Kotui rivers. Following other publications (Mikhin 1941, World atlas 1967 Baranov, Lysyuk, Shurov eds.) it forms by the confluence of Kheta and Kotui rivers, the last river in the upper reaches has a tributary Moyero. In our investigations we keep the opinion by Davydov (1955). According to the data by V. F. Lukyanchikov the sturgeon is distributed in the Khatanga River from its confluence with the Kheta River to the Khatanga Inlet, the upper border of which is the Kresty peninsula. Combining the data by L. S. Berg and V. F. Lukyanchikov (1967) it can probably be assumed that in the Khatanga basin sturgeon is distributed from the mouth of the Kheta River to the Khatanga Inlet inclusive. In the Kheta River, the left tributary of Khatanga, the sturgeon inhabits a range extending 460 km from the mouth to the settlement of Volochanka. Occasionally, it can be found in the lakes in Khatanga floodlands upstream of the Kheta (Kotui according Mikhin (1941)) mouth were it enters with freshets during spring. The main habitat of the sturgeon in Khatanga basin is the middle and upper reaches of the Kheta, 350 km from its mouth.

Anabar River basin. The sturgeon is found in the lower and middle reaches of the Anabar River (Kirillov 1972). However, no recent information is presently available to verify the actual validity of this occurrence. Further studies are needed.

Olenyek River basin. According to some sources the Siberian sturgeon is extremely rare in the Olenek River and only adult specimens can be found in the lower reaches (Lepyoshkin 1966). Other sources indicate that only isolated sturgeon are encountered and not every year, but that they occur mainly up the Pur River and, occasionally, to the mouth of the Chemuudakh River, 1,020 km from the mouth of the Olenyek (Kirillov 1972). Recent information on the occurrences is lacking.

Lena River basin. The northern limit of the Siberian sturgeon in the Lena River is at the mouth of the river, it has occurred in Neyelov Bay (Dryagin 1948a, 1949). During high-water years with large amounts of freshwater inflow, they can be found in Tiksi Bay and its coastal regions of Bulunkan and Sogo gulfs (Kirillov 1950). Within the Lena River, sturgeon can be found up to the village of Korshunovo (Borisov 1928, Dormidontov 1963, Dryagin 1933, 1949, Karantonis, Kirillow and Mukhomediyarov 1956, Kirillov 1972) 1,650 km upstream from Yakutsk (Dryagin 1933). In total, the sturgeon inhabits a section of the Lena River that extents to about 3,300 km in length.

Earlier Dryagin (1948a) has identified the village of Kirensk as the southern limit of the range. In the 1840's, the village of Makarevskoye (Maak 1886), further upstream, was identified as the limit of the sturgeon's range. At the end the nineteenth century, the sturgeon also inhabited the Kirenga River, a tributary of the Lena (Borisov 1928). Thus, during the last 150 years, the southern extension of the range of the Lena sturgeon has been shortened by about 300 km.

The sturgeon inhabits a number of tributaries to the Lena including the Vitim, Olekma, Aldan and Vilyuy rivers (Dryagin 1949, Karantonis, Kirillov and Mukhomedyarov 1956, Kirillov 1972).

In the Vitim River, the sturgeon is found in the lower reaches. Young sturgeons have been encountered in the left tributary of the Vitim, the Nyuya River (Kirillov 1972). According to other observations, the sturgeon occurs in the Vitim up to the mouth of the Tsipa River (860 km from the mouth) (Kozhov 1950), and occasionally even further, up to the mouth of the Kalakan River (Kalashnikov 1978).

Sturgeons have been observed in the Olekma River (Dryagin 1949, Kirollov 1972), but its range in this river was not reported in earlier years. According to our own observations in 1984, the sturgeon does occur in stretches higher than 30 km from the mouth at the confluence with the Chara River, where one 5 year old specimen was captured that was 46 cm in length and 360 g in weight.

In the Aldan River, sturgeons inhabit the lower and middle reaches up to the Ust-Mil settlement. The species was especially abundant in the Amga River, a tributary where its fishery was of significant local economic value (Kirillov 1964). The upstream limit of its range in the Amga River is unknown.

In the Vilyuy River the sturgeon has been found from the mouth to the Vava River and into its tributaries, the Chona, Malaya (Minor) Botuobia and Bolshaya (Major) Botuobia, Tyung and Markha rivers (Kirillov 1972). Before the damming in the late 1960s, it was most abundant at the Chona, Chirkuo and Akhtaranda river mouths. After the construction of the Viluyskoye water reservoir, the environmental conditions changed significantly. Currently, the population in the reservoir has vanished and appears to be very rare in the riverine zone (Kirillov 1986).

Yana River basin. In the Yana River the sturgeon abundance is very low for quite some time (Dryagin 1949). It inhabits the river from the delta to the Verkhoyansk settlement (Kirillov 1972). According to local knowledge, in 1986 it was most commonly found around the Kular (Severnyi) settlement.

Indigirka River basin. In the Indigirka River the sturgeon ranges from the mouth to the settlement of Krest-Maior (850 km from the mouth), and is occasionally found up to Zashiversk (Kirillov 1953, 1972). Recent information on the occurrences is lacking.

Alazeya River basin. In the Alazea River only isolated specimens occur. They have been encountered up to the mouth of a tributary, the Bor-Yuryakh River (Dryagin 1933).

Kolyma River basin. In the Kolyma River, sturgeon have been recorded from the delta up to the Seimchan settlement (Dryagin 1933), i.e. some over 1,500 km upstream, but were primarily recorded up to the Verkhnekolymsk settlement (Dryagin 1948a). According to our observations in 1988, the sturgeons were abundant in the Kolyma River somewhat further upstream, near the mouth of the Popovka River (1,085 km from the Kolyma River mouth). Sturgeons have been recorded in only two tributaries of the Kolyma River, the Korkodon and Ozhogina rivers. They were not encountered in other tributaries of the Kolyma River such as the Omolon, Bolshoi Anyui, Malyi Anui and Berezovka rivers (Dryagin 1948a).

Generalization of data reported above show that the habitat range of the Siberian sturgeon is gradually shrinking (Fig. 35). In the Yenisei River, the construction of the Krasnoyarsk hydroelectric dam shortened the range by about 600 km. In the Lena River, over the past 150 years there has been a reduction of the range of about 300 km.

2. Physical and chemical habitat conditions and anthropogenically induced changes

To understand the potential environmental changes that have occurred and presently are still occurring, we are summarizing in this chapter the available scientific information accumulated during the early and mid decades of the past century while adding here and there more recent information (1980s and 1990s), thereby providing a compiled reference base for comparison when new data become available.

The Ob-Irtysh basin is a habitat of one of the largest populations of Siberian sturgeon. Ob is the largest river in Siberia. The area of the Ob basin is 2,929,000 km^2 and its length is 3,676 km (Davydov 1955). The Ob River originates at the confluence of the Biya and Katun rivers, and is fed primarily by snowmelt. Around Novosibirsk, the proportions of snowmelt, rain, and groundwater sources has been 50%, 27% and 16% with the proportion due to snowmelt increasing with distance downstream. In the upper reaches of the Ob River, there are two characteristic peaks of flow volume, one in late April – early May due to snowmelt, and one in mid-June due to glacial melt. During the first peak, there is a break-up of the ice from the upper reaches through the lower reaches of the river. The maximum rise in the spring water level, up to 13.2 m, is commonly observed at the village Aleksandrovskoye. Further downstream water flow distributes along flood plains, and at Salekhard the water level rises on average not more than about 6 m. There is no summer drought and the water level remains high throughout the summer. For the dynamics within the ice regime of the Ob River it is important to note that the river transports huge masses of warm water from the southern to the northern region, thereby slowing the freeze-up process of the river during the fall. The advance of autumn ice in the lower reaches typically begins towards the end of October (usually the last 10 days) and quickly overtakes all sections of the river, covering the upper (southern) reaches by the end of October. The break-up of ice begins in the upper reaches during the second half of April (usually around the 20th) and occurs in the lower reaches at the beginning of June (Davydov 1955).

Tables 15 and 17 to 20 show that the thermal regime of the Ob River is in principle milder than that of the other Siberian rivers. The annual net heat in the lower reaches of the river at Salekhard is about 1.5 times greater than that in the lower reaches of the Lena River and the Indigirka River. In the middle reaches at Surgut, and at the upper end of the lower reaches at Belogorye, the net heat is roughly twice as great as in the Indigirka River. Consequently, the Ob has a significantly longer period of open waters. At Salekhard the open-water period lasts 146 to 175 days with a mean of 158 days (Votinov 1963).

The waters of the Ob River are weakly mineralized. Following accepted classification they belongs to hydrocarbonate class, calcium group (Alekin 1970). The mineralization decreases from the upper reaches to the lower reaches with a maximum of up to 370 mg/L at Novosibirsk. The net runoff of dissolved substances (total dissolved solids) in the Ob River is large and amounts to 32,800,000 tonnes per year (Davydov, 1955).

Of particular importance to the lower Ob is the annual winter hypoxia when the river is fed by ground- and marsh waters, which, in passing through the ground, rich in decomposing organic matter, lose much of the dissolved oxygen to the oxidation of organic matter. Thus, these waters are almost entirely anoxic when entering the river system. Furthermore, large quantities of humic substances and ferro-salts entering the river consume a lot of the oxygen, resulting in oxygen concentrations as low as 5% saturation, and in some parts of the Ob (e.g. the mouth of the Vakh) to 0% oxygen by the end of winter (Votinov 1963). These hypoxic and anoxic zones occur within a 2,400-km reach from Kolpashevo to the mouth of the Ob, and in the upper part of the Ob Bay. Hypoxia does not occur everywhere simultaneously but these events begin in the Surgut Ob, where the oxygen-deprived waters from the Bolshoi (Great) Yugan, Vakh, Agan and other tributaries enter the Ob. Over time the hypoxic front advances upstream to the mouth of the Tym River, and hypoxic events occur between the Tym and the Ket River, however, not every

Table 15:
Mean monthly temperatures (°C) in the lower reaches of the Ob River (Votinov 1963).

Location, settlement	Month						Net Annual Heat (degree days)
	May	June	July	August	September	October	
Salekhard		9.34	15.74	15.24	9.03	2.33	1584
Berezovo	4.20	12.0	16.90	16.90	9.70	3.20	1928
Belogorye	6.16	15.20	19.39	17.68	11.48	4.12	2269
Surgut	4.45	14.10	18,97	17.36	11.26	3.26	2126

Table 16:
Pollution levels in the Ob-Irtysh basin in terms of exceeding the maximum concentration limit (MCL) in 1990 (Annual Report on Surface Water Quality of USSR, 1991).

Location, settlement	Petrochemicals	Phenols	Copper complexes	Zinc complexes
Ob (Fominskoye - Aleksandrovskoye)	1-43	2-24		
Ob (Nizhnevartovsk - Salekhard)	6-42	1-20		
Irtysh (Buran - Bobrovskoye, Kazakhstan)	2-12		3-8	0-4
Irtysh (Cherlak – Khanty-Mansiysk)	7-43	5-16	5-16	
Bukhtarma			7-20	
Bukhtarminskoye reservoir	2-9		4-10	0-2
Ust-Kamnogorskoye reservoir	3-5		4-8	

Table 17:
Mean monthly temperatures (°C) in various sections of the Lena River (Sokolova 1951).

Location, settlement	Month						Net Annual Heat (degree days)
	May	June	July	August	September	October	
Kirensk	3.0	12.6	17.1	15.6	8.9	1.9	1811
Olekminsk	3.0	13.7	19.0	16.9	10.2	1.6	1973
Pokrovskoye	1.7	13.1	18.7	16.7	9.3	1.5	1869
Zhigansk		12.0	17.0	15.2	8.6	1.0	1647
Kyusyur		5.3	14.2	12.6	6.8	0.8	1218

Table 18:
Mean monthly temperatures (°C) in the Lena River near the Tit-Ary settlement, measured between 1981 and 1983.

Month	June	July	August	September	Net Annual Heat (degree days)
Temperature	3.8	12.3	13.0	5.8	1079

Table 19:
Mean length of duration of open water in various sections of the Lena River (Shostakovich 1909).

Location	Days
Kachug	184
Kirensk	164
Vitim	181
Olekminsk	172
Yakutsk	155
Bulun	138
River mouth	100

Table 20:
Mean monthly temperatures (°C) in various sections of the Aldan River (Sokolova 1951).

Location	Month					
	May	June	July	August	September	October
Uchur	1.3	10.2	16.8	13.4	7.3	1.1
Ust-Mil	1.9	11.8	17.3	15.3	9.3	1.4
Krest-Khaldzhai	1.4	11.4	17.4	15.8	9.5	1.7

year. The hypoxic front advances downstream at a mean speed of 30 km a day, reaching the mouth of the Irtysh River by the end of December or beginning of January and the Salekhard city one month later. The hypoxic front reaches the mouth of the Ob River and it's delta much later, typically at Novyi Port during the middle of May (Votinov 1963).

Hypoxic events on the Ob exert a great influence on the life of fish as they are forced to either swim downstream ahead of the rapidly advancing hypoxic front or migrate upstream to stay alive above the hypoxic zone. A few fish remain near well oxygenated springs and tributaries (Votinov 1963).

The end of the hypoxic event occurs in spring with the flow into the river of oxygen-rich meltwater advancing in the same sequence as the hypoxic zone. The Taz and Pur rivers entering the Ob Bay are also hypoxic in winter (Votinov 1963).

Irtysh River. The Irtysh River is the largest tributary of the Ob River. It originates in the Mongolian Altai Mountains on Chinese territory. The Irtysh River is 4,422 km long and the area of its drainage basin covers 1,595,680 km^2. The average annual discharge is 89.3 km^3 (Domanitskiy et al. 1971). From its origin to Lake Zaisan (currently a part of the Bukhtaminskoye Reservoir), the river is known as the Chorniy ("Black") Irtysh; down to the Altai foothills it is known as the Beliy ("White") Irtysh (Davydov 1955). Downstreams Omsk city, the Irtysh traverses swampy country and it's right tributaries, the Tara, Uy, Turtas, Demyanka and others, have been rich in humic substances (Vasiliev et al. 1957). The hydrography of the Irtysh River, therefore, varies in different reaches. The upper reaches have a dramatic spring freshet, whereas in the steppe zone the spring water level rises more gradually with peak in late May or June. In the taiga zone, there is a further delay in the timing of the peak and the freshet extends into the summer (slower snow melt in the forests and buffering in marshes) (Davydov 1955). In the lower reaches of the river, typical annual water-level oscillations are in the order of 8-9 meters, with maxima around 14-m. The regulation of the flow regimes in the upper reaches due in part to the construction of the Ust-Kamennogorskaya hydroelectric station (HES), has had practically no influence on the overall hydrography of the lower Irtysh (Vasiliev et al. 1957).

The freeze-up of the lower reaches of the Irtysh River occurs usually in the beginning of November and in the upper reaches in the middle of November. The break-up of the ice in the upper reaches occurs in the middle of April, and in the lower reaches at the end of April or beginning of May (Davydov 1955). The hypoxic event (so distinctive in the Ob River) occurs in the Irtysh River only in its lower reaches up to the mouth of the Konda River, 112 km from the mouth of the Irtysh and more rare to the mouth of the Demyanka River, about 340 km from the mouth (Votinov 1963).

The environment for Siberian sturgeon in the Ob-Irtysh basin due to the evolving economic activities were gradually but significantly changed. The main changes were caused by dam construction (Figure 35) and water pollution. Dam construction influenced the upper reaches of Ob and Irtysh rivers.

Between 1957 and 1959, the Novosibirsk reservoir, with a surface area of 107,000 ha, was created. The flooded zone stretches for 203 km from the settlement of Nizhniye Chemy to the town of Kamen-na-Obi (Isaev and Karpova 1989).

Before flow regulation, there were two principal spawning grounds in the Ob River: the lower grounds, stretching for 782 km from the settlement of Allak to the mouth of the Chulym River, covering 35,100 ha, and the upper spawning ground stretching from the confluence of the sources (Biya and Katun) to the mouth of the Charish River, and likewise in the Katun and Anuiy rivers, stretching for 200 km and covering 8,200 ha (Petkevich 1952). The dam at the Novisibirsk HES has no accommodations for passing fish (Kirillov 1991) and completely blocked the access to the entire upper spawning ground and part of the lower spawning ground (blocking about 40% of the total spawning grounds), as well as 50% of the overwintering habitat of the so-called semi-anadromous form of the Ob sturgeon (Votinov 1963, Kirillov 1991). As a result, the reproduction of this form in the first years after the construction of the dam was reduced by a factor of 8-10 times (Votinov 1963). In 1961-1965, it was lowered by a factor of 12-180 times compared to the 1955 spawning activity. Between 1965 and 1969, the seaward migration of juvenile sturgeon increased somewhat, though it remained 5-6 times below pre-dam levels. In 1971, the number of migrating juveniles increased significantly due to a strong year-class in 1969, but on the whole the reproduction of sturgeon has fallen by a factor of 5-6 in comparison to the period prior to flow regulation (Votinov et al. 1972). The effect of the Novosibirsk Dam on sturgeon reproduction is not limited to the reduction of the spawning areas alone. An analysis of the dynamics of sturgeon juvenile populations and the accompanying water levels has shown that the most abundant generations appears in years with high water level. Currently, during the spawning season during May-June, the reservoir is filled to compensate for winter drawdown. As a consequence, the spring flows downstream of the dam are significantly reduced compared to previous years that preceded the dam's construction (Votinov et al. 1975). The drop in reproduction of the Ob sturgeon by 1963 was due not only to the dam, but also to the blockage of spawning grounds in the Chulym River due to timber-rafting and loss of spawning grounds below the mouth of the Tom River caused by industrial pollution (Votinov 1963). Thus, the construction of the Novosibirsk dam also promoted the reduction of the habitat for the "semi-anadromous" population and subsequently decreased its reproduction level. Upstream of the dam, a resident form of the Ob sturgeon has emerged. The number of young sturgeon in the reservoir grew steadily for several years. Up to 1976 the numbers of sturgeon seem to have stabilized, and this was represented primarily by immature individuals less than 13 years of age, however, the number of spawners was small. The growth of sturgeon in the first years of the existence of the reservoir improved dramatically (Setsko 1976). Juvenile sturgeon above the reservoir are characterized by a slower growth rate than in the reservoir (Setsko 1976), but, in general, they grow faster than before the river was regulated (Solovov 1997).

The habitat quality for sturgeon spawning in the upper reaches of the Ob River has worsened since the construction of the Novosibirsk HES. Because of flooding and other effects, the total area of the spawning grounds has been reduced by a factor of about 10 times from originally 8,200 hectares to about 850 hectares. In the lower reaches of the Katun River, the mining works at the Talitsa deposit of gravel located in the river bed reduced the area of the spawning grounds also by a factor of about 10 times from 2,500 ha to about 200-300 ha. The spawning grounds near the confluence of the Biya and the Katun rivers are no longer of any significance due to thermal and chemical pollution from the Biysk thermoelectric plant (Solovov, 1997). The range of the sturgeon in the southern part of Ob basin has also been reduced (see above).

The effect of dam construction on sturgeon reproduction and survival in the Irtysh River are highly significant. In the middle and lower reaches of the Irtysh River, the spawning grounds have not been well studied, however, they have been noted (a) near the village of Suzgun located about 12 km below Tobolsk city (640 km from the mouth) (Votinov 1963), (b) near the settlement of Abalak (689 km from the mouth) and (c) in a tributary to the Irtysh, the Tobol River, 30 km from it's mouth (Kasyanov et al. 1979). In the upper reaches of the Irtysh River from Semipalatinsk upstream to Lake Zaisan, there were once broad spawning areas for sturgeon before the beginning of the dam construction. Sturgeon also spawned in the lower reaches of several of the Irtysh's tributaries – the Ulba, Uba, Bukhtarma and Kurchum rivers.

In the upper reaches of the Irtysh River, the Bukhtarminskoye (1960-1964), Ust-Kamenogorskoye (1950-1954) and Shulbinskoye (1987) reservoirs were constructed at km 3,262, 3,162 and 2,993 upstream from the river mouth, respectively. The 510-km long Bukhtarminskoye reservoir includes Lake Zaisan and the delta of the Chorniy (Black) Irtysh. The Ust-Kamenogorskoye water reservoir is of the canyon-type (Isaev and Karpova 1989).

The information on the main spawning grounds for Irtysh sturgeon is somewhat more recent than those reported above for other river systems. In the latter part of the last century, they were once located between the mouth of the Shulba River and Ust-Kamenogorsk (*Fish of Kazakhstan* 1986), and have been inundated by the Shulbinskoye reservoir. The section of the river between Ust-Kamenogorsk to the mouth of the Bukhtarma, once known as the Bystriy ("Fast") Irtysh and currently within the Ust-Kamenogorskoye water reservoir, served as the main wintering grounds for sturgeon. Extensive spawning grounds were located in the section of the Irtysh River between the mouth of the Bukhtarma River to Lake Zaisan (the Beliy Irtysh). Here, these grounds stretched in a continuous chain over roughly 70 km from the mouth of the Bukhtarma River to the mouth of the Bolshaya Bukon' River, after which they continued intermittently to Djambai. The most well known spawning grounds here were found near the Gorshechniy islands, the Gladkovskiy sandbars, and the villages of Ust-Chernovaya, Tassaut, Kanonerka, Baty, and Djambai. All of these spawning grounds are now in the deep water zone of Bukhtarminskoye reservoir and can no longer serve this purpose (Votinov et al. 1975, *Fish of Kazakhstan* 1986).

The construction of the Ust-Kamenogorskoye and, especially, the Bukhtarminskoye reservoirs has led to a dramatic cooling of the waters (Votinov 1965). The temperature of the water in the Irtysh branch of the Bukhtarminskoye reservoir at a depth of 20 m does not exceed 9-10°C and in deeper waters remains around 4-8°C (Ereschenko 1966a). Closely correlated with temperature is the amount of benthic biomass, which is greatest at depths of 10 m but is much lower at depths exceeding 20 m. In the most productive and shallow lake section of the reservoir (Lake Zaisan), the benthic biomass approaches 430 kg/ha. In the deep-water mountain/valley section of the reservoir, it is approximately 182 kg/ha. The average benthic biomass in the reservoir was approximately 350 kg/ha at the time of study (Kozlyatkin et al. 1973).

The cooling of the waters as a result of hydroelectric development and worsening of sanitary conditions (release of untreated wastes) in the Irtysh River below the Ust-Kamenogorsk had effectively destroyed the viability of all spawning grounds in the area of the Shulbinskoye water reservoir even before its construction (Ereschenko 1966a and b, *Fish of Kazakhstan* 1986).

Thus, the massive development of hydroelectric power dams has practically completely destroyed the entire main spawning grounds of the "semi-anadromous" form of sturgeon in the upper Irtysh River. There remain only some smaller spawning grounds of the resident Zaisan- Chorniy Irtysh population. These are distributed in the Chorniy Irtysh River, mostly on Chinese territory and in the Kurchum River (Votinov 1965, 1966, Votinov et al. 1975, *Fish of Kazakhstan* 1986).

Currently, the Ob-Irtysh basin suffers from pollution of large industrial projects, including oil and natural gas extraction, forestry, transportation, military-industrial activities, and other developments. For more than 35 years Siberian sturgeon populations are strongly affected by this development. At the end of the 1950's and early 1960's the potential effects of this development were not obvious and no pathologies were observed in the reproductive systems of the Ob sturgeon population (Votinov 1963). However by 1995 sexually mature individuals exhibited clearly a wide spectrum of pathological features in the gonads, up to total sterility (Ruban 1996). From 1981 to 1996, abnormalities in gonads became noticeable in peled (*Coregonus peled*) and muksun (*Coregonus muksun*) in the lower Ob populations during the foraging period. An increase in hermaphroditism was also observed in these species. Specimens of dace (*Leuciscus danilewskii*), roach (*Rutilus rutilus*), ide (*Leuciscus idus*), and perch (*Perca fluviatilis*) exhibited abnormalities in oocyte nucleii, showing cytolysis and atresia during stages of cyto- and trophopasmic growth, a degeneration of the gonads in females, and resorption of spermatocytes as well as malformed spermatozoids in males (Selyukov 1997).

Oil extraction in the Ob basin began in the late 1950's and early 1960's. The first 2,000 tonnes of West Siberian oil was delivered on the ship *Fersman* on May 26, 1964, to Omsk from the Ust-Balyksk oilfields (Tonyaev 1972). By 1984, oil production in the Tyumen region had reached one million tonnes per day. The development of the oil and gas industries was accompanied by intense pollution of the Ob basin with oil and oil by-products, the concentrations of which in the waste waters of oilfields attained 0.2-80 g/L, including phenols, synthetic surfactants, heavy metals, chlorides, sulphates, diethylene glycol, methanol and other substances. In the acquisition of oil, over 150 chemicals are used (Mikhailova 1991). Of all the oil related pollutants in the water bodies in the Tyumen region, 80% to 85% are petro-hydrocarbons. The principal sources of oil-related pollution are accidents in the pipelines and storage pits for flares, accelerants, fuel and lubricants and industrial waste runoff. Expert estimate the quantity of oil products entering the Ob-Irtysh basin in the order of 112,960 tonnes per year (Mikhailova 1991). The contaminants in the water derived from oil are highly resistant to chemical and biochemical degradation (Gasanov 1983, Novosadova 1985) except the highly volatile fractions. The maximum level of purification does not exceed 85%, and hydrocarbons undergoing biological purification remain conservative toxins once they enter the water (Rogovskaya 1967, Mikhailova 1991). Especially hard hit are smaller rivers of the Ob basin. Thus, for example, a one-time discharge of waste waters from oilfields in the Surgut and Nizhnevartovsk regions into the Kalinovaya and Vatinskiy-Yegan rivers changed drastically the salinity (Mikhailova 1991). During 1950 to 1985 the chemistry of the waters in the Ob-Irtysh basin was radically transformed as a result of industry. The mineral content of the water has increased 1.5 times in the low-water period and 2.5 times during the high-water period to 154-235 mg/L and 117-128 mg/L, respectively (Smirnov 1975, Mikhailova et al. 1988). In lower reaches of the Ob permanganate oxidization demand was 2-3.5 times higher than in upper reaches, ammonia nitrite increased by a factor of 4-6 and nitrogen nitrite by a factor of 1.5-2.0.

Several tributaries of the Ob – the Vakh, Bolshoi (Great) Yugan, Pim, Salym, and Vatinskiy Yegan – traverse the oil-producing regions and, polluted with petrochemicals, enter the Ob River and in its middle reaches. As a result the concentration of petrochemicals in water exceeds the maximum tolerance limit (MTL) by a factor of 4-6 (Mikhailova 1991). Because these chemicals are highly insoluble, their ambient concentration generally does not exceed 0.2-0.5 mg/L. Oil, as a surface film and suspended in the water, is transported by the Ob River into the Arctic Ocean at a rate of 120,000 tonnes annually. Roughly 30-50% of the heavier components of the oil settle to the bottom of the river (Voroshilova and Dianova 1960, Izyurova 1955). The oil that enters the grounds at river bottom then becomes a source of secondary pollution. The accumulation of oil components in the bottom grounds negatively impacts the microflora and correspondingly processes of destruction. Since the 1960, the concentration of oil components in the bottom grounds of the lower Irtysh River has increased by a factor of 2,4, and in the middle Ob by a factor of 45, attaining levels over 1 g of oil fractions per 100 g of sediment at the beginning of the 1990ies (Mikhailova 1991).

There are several ways in which oil pollution and its various breakdown products cause direct harm. Firstly, there is the effect on the nervous system of the fish. In the initial stages of intoxication, the accumulation of the low molecular weight oil products in the brain is accompanied by a narcotic effect, followed by a depression of the respiratory centre, weakened vision, a disruption of spatial orientation, and ultimately death (Woodward *et al.* 1983). Secondly, with accumulation of hydrocarbon compounds the fish is no longer suitable for human consumption. Finally, the trophic capacity of the water body is greatly reduced as only opportunistic species will be able to survive and thrive. As an example of the more indirect influences of oil pollution, it has been shown that the accumulation of petrochemicals in the reproductive tissues of fish leads, in part, to (a) a massive degeneration of oocytes in the pre-vitellogenesis stage of development as observed in the Ob muksun (*Coregonus muksun*) (Selyukov and Stepanov 1988), (b) hatching of largely unviable larvae and (c) a sharp rise in malformation rates as observed among the peled (*Coregonus peled*) (Bogdanov 1983). Oil pollution in water bodies also causes changes in the plankton and benthos communities. In the latter, a decline in biodiversity, a reduction in the range of dominating species and their replacement by carnivorous *chironomidae* larvae as well as the reduction in whole numbers of organisms and dominating of small size species of *chironomidae*, oligochaetes and molluscs has been observed (Yukhneva 1970, Mikhailova 1991).

The middle and lower sections of the Ob basin are also under the influence of pollution caused primarily by the extraction of oil and natural gas. However, the region is also the place where in the past the remains and stages from rockets launched from the Baikonur and Plesetsk launch sites are deposited, which pollutes the natural environment with heptyls and other harmful substances. As a result, in 1994 the waters of the Tobol, Tom, and Chulym rivers were categorized as extremely polluted and unusable for drinking water utilization (Rikhvanov 1996). In 1991 in the Tom River near the city of Kemerovo the following highly toxic substances were identified: aniline, carpolactam, formaldehyde and methanol. In the Iset River below Yekaterinburg, mean annual concentrations of petrochemicals, ammonium nitrite and copper and zinc complexes exceeded the MCL values more than tenfold (*Governmental Report on the Conditions of ...* 1992).

In the upper and middle reaches of the Ob-Irtysh basin there are numerous chemical and petrochemical industries, the significant rate of pollution caused by these industries in Kazakhstan in the cities of Pavlodar and Ust-Kamenogorsk (Rikhvanov 1996) is obvious. Around Novosibirsk, the Ob River is polluted by runoff from the industries belonging to Ministry of power engineering, Ministry of metallurgy, Ministry of municipal economy and, most of all by runoff from the armaments industry. Thus, as a result of a one-time pollution discharge (equivalent to a chemical spill) from Novosibirsk factories in 1990, the concentration of phenols in the Ob River grew temporarily to 88-92 times the MCL values (0.088-0.092 mg/L). The average annual concentration of phenols in 1990 below the Novosibirsk hydroelectric dam stood at 23 times the MCL and in the region near Aleksandrovskoye village in 1989 and 1990, at 120 and 43 times the MCL, respectively. The principal polluters are the pipelines of the oil infrastructure, shipping, and timber rafting (*Annual Report on Surface Water Quality in the USSR [ARSWQ]* 1991). Shipping has been extensively developed over the past several decades, with 700 new vessels added to the Ob-Irtysh fleet between 1975 and 1980. The volume of goods transported by water is rather high; 3,850,000 tonnes of timbers alone were transported in rafts on the upper and middle reaches of the Ob River in 1980 (*ARSWQ* 1991).

The summary data on the level of pollution of the Ob-Irtysh basin by some toxic substances is given in Table 16.

From Table 16 it can be seen that the primary components of pollution before and during the early 1990ies in the lower and middle reaches of the Ob River were petrochemicals and phenols, whereas the Irtysh is polluted primarily by petrochemicals and copper and zinc complexes. In some years, the amount petrochemicals and phenols exceeded the MCL reached 120 and 131 respectively (*ARSWQ* 1991).

In past years, the section of the Ob River from Nizhnevartovsk to Salekhard contained concentrations of petrochemicals and heavy-metal ions that exceeded the allowable MCL for the fisheries water bodies. However, there has been an apparent decline in petrochemical concentrations. Between 1986 and 1990 their concentrations were in the 0.31-0.62 mg/L range, whereas in 1995-1996 the concentrations dropped to 0.19-0.26 mg/L. In flood water bodies in the Surgut and Nizhnevartovsk regions, 3-6% of the bottom grounds are considered extremely polluted, 14-18% are considered moderately polluted and 29-37% are considered weakly polluted (Andrienko et al. 1997).

At the mouth of the Ob River and in the Ob and Taz bays, petrochemical pollution has declined in recent years, but phenols, pesticides, heavy metals and synthetic surfactants remain in excess of the maximum concentration limit (MCL) set for fisheries water bodies. According to ecological-sanitary indices they belong to the weakly and moderately polluted class of water bodies. In terms of petrochemical concentrations in the bottom sediments, the lower reaches of the Ob River are classified as weakly and moderately polluted, but the concentration of heavy metals exceeds the MCL for soils. The waters of the lower Ob River are classified as moderately polluted according to the hydrobiological indices (Semenova et al. 1997).

The most polluted tributary of the Ob River is the Tom River, draining the highly industrial regions of Kuzbass - Mezhdurechensk, Novokuznetsk and Kemerovo. In certain sections of the Irtysh, petrochemical, copper complex, and phenol concentrations can exceed 123, 66 and 77 MCL's, respectively (*ARSWQ* 1991). These elevated quantities of heavy metals are dangerous for fish not only by direct biological interaction, but also in their transfer and accumulation into the trophic chain. It is known that bioaccumulation of manganese, cobalt and copper by molluscs leads to an increase in the concentrations of these substances by factors ranging from 2 to 1800 (Shakhmayev 1973). This, in turn, increases their accumulation potential in the benthophage organisms, consequently including the Siberian sturgeon.

On the whole, it can be said that pollution affects the ichthyofauna of the Ob basin substantially. It is primarily evident in the right tributaries of the middle Ob, where even a reduction in the intensity of the local fisheries (e.g. for pike, roach, dace, and ide) has not been able to increase the rate of older cohorts in populations due to the high rates of mortality affecting the older year classes (Andrienko *et al*. 1997).

Yenisei River. The Yenisei River is the largest river by volume in Siberia. The basin encompasses an area of 2,599,000 km^2 and the length of the river is 3,354 km. The Yenisei River originates at the confluence of the Biy-Kem and Kha-Kem rivers. Snowmelt provides somewhat below 50% of the total volume, while rain, at the settlement Oznachennoye and the city of Krasnoyarsk, provides about 36-38%, of total runoff and groundwater provides roughly 16% of the water supply to the system (Davydov 1955). In the upper reaches, the spring freshet is extended but is less dramatic than in the lower reaches, where the peak is higher though it's duration is shorter. The freshet moves downstream twice faster than spring comes. Thus, break-up of the ice is due to rising water, which leads to the formation of ice jams. In the middle reaches, the water level peaks occur in the middle of May while in the lower reaches they appear in the middle of June. A characteristic feature of the Siberian rivers that flow north into the Arctic Ocean is that they carry an enormous mass of heat, delaying freeze-up in the fall. The Yenisei River freezes about 10 days later than it's tributaries. The autumn ice drifting appears in the first ten days of October. The break-up of the ice on the Yenisei is violent, accompanied by ice jams, and extends over a month. Break-up begins in the first third of May in the upper reaches and continues into the first third of June in the lower reaches. The turbidity of the water is weak, 19-50 g/m^3. The waters of the river are weakly mineralized (less than 300 mg/L) and belong to the low level hydrocarbonate class (Davydov 1955).

The Sayano-Shushinskoye Reservoir, which is 1-7 km wide and up to 220 m deep, has flooded a 320-km long section of the Yenisei River (2959-3279 rkm from the mouth) since 1978 (Isaev and Karpova 1989).

The Krasnoyarskoye Reservoir, built in 1967-1970, flooded a stretch of 338 km of the Yenisei River. The reservoir is 2.5 to 15 km wide and has a maximal depth is 105 m with an average depth of 36 m (Isaev and Karpova 1989).

Angara River. The Angara River is the largest tributary of the Yenisei River. In the Baikalo-Angara basin, 56% of the total area is within the Lake Baikal basin and 44% drains to the Angara basin. About 47% of the volume of the Angara comes from Lake Baikal. Before the regulation of the river discharge by dam construction, the maximal water level occurred at the end of May and remained high through the summer until October. Low water levels were typical of April (Davydov 1955). Before flow regulation, the unique feature of the Angara, related to the influence of Lake Baikal, was that the freeze-up in the upper reaches did not occur until the middle of January, 1.5 months later than in other rivers. In the lower reaches, the freeze-up occurred in the normal period. Ice jams, characteristic of all Siberian rivers, causes winter floods in the Angara. The break-up of the ice in the Angara was a slow process, occurring in the first third of April at Irkutsk and in the middle of May in the lower reaches. The spring freshet is weak. The water of the Angara is weakly mineralized (90-150 mg/L) and of the hydrocarbonate class. Turbidity is also weak, around 24.8-142 g/m^3 (Davydov 1955).

Lake Baikal. Located 454 m above sea level, Lake Biakal is 636 km long and 79.4 km wide at its widest point. Its surface area covers about 31,500 km^2 and its volume is estimated with 22,160 km^3. The drainage basin is 557,500 km^2, of which 83.4% or 419,940 km^2 is within the basin of the Selenga River (Votintsev 1969). An important feature of the lake with respect to fish habitat, including the Siberian sturgeon, is its great depth with relatively small amount of shelf waters along the littoral zone. The lake depression can be subdivided into three distinct hollows. The southern hollow, located to the south by the mouth of the Selenga River, has a maximum depth of 1,473 m and an average depth of 810 m. Its area is 6,890 km^2. The middle hollow takes up that part of the lake from the Selenga shallows to the northern end of the Olkhon Island. Its maximum depth is 1,741 m with an average depth of 803 m and it covers an area of 11,295 km^2. The northern hollow is located to the north of the northern end of the Olkhon Island. Its maximum depth is 938 m with an average depth of about 564 m and it covers 13,315 km^2 (Votintzev 1961). Underwater ridges separate the hollows from each other. The ridge that separates the southern and middle hollows is referred to as the Selenga shallows, the depth of which does not exceed 428 m. The middle and northern hollows are separated by the Academicheskiy ridge and its depth does not exceed 400 m. The largest of the 336 rivers that feed Lake Baikal are the Selenga River, the Verkhnyaya (Upper) Angara River, and the Barguzin River. These rivers, provide 50, 13-14 and 9% of inflow from surface water by volume, respectively.

The freeze-up of the Baikal extends from the middle of December in the northern end to the end of December or early January in the southern end. The break-up of the ice occurs at the end of April or in early May in the south, while in the north, the break-up occurs about a month later (at the end of May or in early June). The thermal regime of the lake is complex. The dynamics of the warming and the cooling of the deep and surface layers are significantly different. We will restrict ourselves to a discussion of the temperature regime up to 200-250 m depth, since the Baikal sturgeon is encountered at the depth not more than 200m (Yegorov 1961). After the break-up of the ice as a consequence of the intense mixing of water masses, the temperature of the water rises very slowly and complete homothermy in the upper layer up to the depth of 250 m happens only in the end of June at a temperature of 3.6-3.8°C. The warming of the surface waters of Baikal in summer occurs slowly. Maximum temperatures are observed at the end of July or in early August, occasionally even in the beginning of September.

The cooling of the waters in the fall is influenced to a large degree by the winds, which move the warm surface waters to the south-eastern shore. The autumn homothermy in the surface layer occurs in the second half of November at a temperature of 3.6-3.8°C (Votintzev 1961).

The waters of the Baikal are weakly mineralized and of the hydrocarbonate class, calcium group. In terms of the content and relationship of the main components of the ionic content, and the dynamics biogene substances and organic substances content, the lake has no analog system. The closest comparison to the chemistry of Lake Baikal are the waters of the Great Lakes in North America, but these latter contain greater iron and silicic acid concentrations (Votintsev 1961). Currently, the lake is undergoing significant changes mainly due to pollution from industrial runoff, the contribution of which in separate years by volume is about 230.45 million m^3, of which polluted waters compose 169 million m^3 (*Government Report on the Condition of the Environment in the Russian Federation [GRCE]* 1992).

The Siberian sturgeon, as noted above, was rather rare in the Yenisei River above Krasnoyarsk (Podlesnyi 1955). Since the early 1990ies, they are no longer observed above the Krasnoyarsk dam, built in 1967 (Isaev and Karpova 1989, personal communication Yu. V. Mikhalyov). Therefore, we will not examine changes in the hydrologic and other conditions in the sections of the upper Yenisei River that fall into the Krasnoyarskoye and the Syano-Shushinskoye reservoirs.

The largest changes related to hydroelectric construction projects occurred in the Angara, which has been transformed into a series of reservoirs, practically from its source to its mouth, over a net length of 1307 km.

The Irkutskoye reservoir is of the channel type, built in 1956, 65 km below the source of the Angara (Askhaev and Gomenyuk 1966). Its area is 190 km^2, of which 28% corresponds to the Angara channel itself, 8% represents flooded islands, 64% appear as drowned land formerly used as fields, gardens, railways, settlements, marshes, meadows, brush, and cultivated mixed and coniferous forest. The length of the reservoir is 65 km; its widest point is 3.5 km; its depth is 7 m in the upper reaches and 35 m at the dam. The oxygen regime is favourable, with saturation levels oscillating between 71% to 125% over the year both on the surface and the bottom layers. The reservoir is covered by ice from the end of November – early December to early –mid May (usually 5-20). The upper reaches of the reservoir do not freeze over. The mean residence time of the water in the reservoir is roughly half a month. The reservoir was filled to the project mark in August 1958. The filling was accompanied by the disappearance of rheophilic forms of mayflies and midges (*Tendipedidae*) from the benthos of the Angara. Lake Baikal influences the reservoir in that there is a well-expressed current, a weak warming in the summer, high clarity, and low mineralization. After the filling of the reservoir, the benthic biomass rose to 225 kg/ha, of which midge larvae made up 66.8% (Golyshkina 1967, Yerbaeva 1967, Tugarina *et al.* 1967, Tugarina and Gomenyuk 1967, Isaev and Karpova 1989). Later, the benthic biomass dropped to 101 kg/ha. The percentage of midge larvae dropped to 43.2% in 1960 and later rose to 89-94% (Yerbaeva 1967). Eventually, the benthic biomass in the channel part of the reservoir was 145 kg/ha, made up of 64% oligochaetes and 27% midges. According to limnological indices and trophic conditions, the Irkutskoye reservoir is comparable to oligotrophic water bodies, such as the bays of Lake Baikal (Yerbaeva 1973). In 1963-1964 young Baikal sturgeon, 0.75 to 2.5 kg in weight, were observed in the reservoir (Tugarina and Gomenyuk 1967).

Before the construction of the Bratskoye reservoir subsequent timber rafting destroyed the significance of the Angara River for fisheries. Beyond the sedimentation of bark, branches and woodchips, much damage was caused by braking chains on the rafts which tore up the bottom of the river, destroying the spawning grounds and wintering holes (Ramazanov 1966).

The Bratskoye reservoir, built in 1966, flooded portions of three rivers; 565 km along the Angara, 360 km on the Oka, and 100 km along the Iya. Its surface area is 547,000 ha. The residence time of the water is roughly two years. The average depth is 32.6 m, its greatest depth is 106 m, and its greatest width is 33 km. Zones 0-10 m in depth make up 21.4% of its total area. Its surface waters are warmed in the summer to 19-21°C, and deeper waters to 15-17°C (Lukyanchikov 1966, Shulga 1967, Isaev and Karpova 1989). In contrast to the oligotrophic Irkutskoye reservoir, the Bratskoye reservoir is characterized by intense eutrophication. In the central channel, the benthofauna is dominated by oligochaetes (82% of the biomass) and in the littoral zone by midges (69%) (Yerbaeva 1973). Siberian sturgeon in the reservoir are distributed over the entire area, up to the upper reaches of the Oka and Angara rivers, and also enter the Belaya and Irkut rivers. They are encountered to a depth of 10 m, but can be found up to 20-25 m, preferring areas above drowned islands not littered with forests and brush. Significant numbers of sturgeon are concentrated around the wintering holes in the middle reaches of the Oka River (Lukyanchikov 1967b). Since 1962, young Baikal sturgeon have been introduced in the Bratsk reservoir from the Belskyi Hatchery of the Irkutsk region. Between 1976 and 1985, 0.6 million juvenile Siberian sturgeon 3 to 8 g in weight were released in the reservoir. In the summer period, the juveniles remained close to the mouth of the Belaya River near the location of the release. Between 1984-1985, rather large individuals were caught, including females 140-147 cm in length and 19-22 kg in weight and males 95-132 cm and 7.5-16 kg in weight. These captures were in the same location for hatchery spawning purposes (Afanasyev and Polyakov 1986).

The Ust-Ilimskoye reservoir, built in 1974-1977, has a river-like form. Its length along the Angara River from the dam of the Bratskoye reservoir is 302 km and 300 km along the Ilim. Its area is 188,000 ha and it has a mean depth of 32 m with a maximum of 92 m. The residence time of the water is between 6-8 months. 69% of the reservoir floods forest and shrubbery. Due to the reduced current velocity, the open (ice-free) period of the reservoir has shortened to 150 days (Isaev and Karpova 1989).

In the first years of its existence, the flooding of a virgin forest less than 12 cm in thickness, soil cover, moss marshes containing peat caused a rapid drop in oxygen levels and an elevation of carbon dioxide. In 1977, the bottom layer oxygen content was a mere 0.1% of saturation in the zone of the Bratsk Forestry Complex waste water runoff. The surface concentration was at

38% of saturation. The lower layers of the water were contaminated with hydrogen sulfide, the concentration of which attained 30 mg/L. On the whole, there was a net reduction of oxygen concentration by 13-18%. Above the flooded agricultural fields, there was no dissolved oxygen but there was a high concentration of hydrogen sulfide (Seryshev and Ozherelyev 1978). Near zone of the Bratsk Forestry Complex waste water runoff, the benthic biomass was 0.932 g/m^2 and was dominated by the oligochaete *Tubifex tubifex* (Romanenko and Tryamkina 1978).

The flooding of the Boguchanskoye reservoir, the dam of which is located in the lower reaches of the Angara River, began in 1987. It is 375 km long and has a surface area of 232,600 ha. Its maximum width is 14-15 km; its minimal width is 1.2 km. It has an average depth of 25 m with a maximum of 70 m. As a consequence of the transformation of the hydrological conditions in the flooded zones, the spawning grounds of the Siberian sturgeon, sterlet and several other species have lost their significance (Isaev and Karpova 1989).

For the Angara River, as with other rivers in Siberia, there has been a significant drop in acipenserid populations (sturgeon and sterlet). Harvest, which was 440 tons in 1897, had dropped to 72 tons by 1928, and oscillated between 12 and 67 tons during 1941-1947 (Shumayilov 1986).

The construction of a series of reservoirs in the upper Angara River (Irkutskoye, Bratskoye, Ust-Ilimskoye) and large-scale timber rafting has transformed the conditions of existence for fish, including sturgeon. Hydroconstruction has led to a drop in the heat flow of the river. The Bratsk HES alone has lowered the heat flow from 888.3 billion gigacalories to 633.9 billion gigacalories. The construction of the dams on the Angara River caused the disappearance of freshets, water level is now determined by releases through the dams (Shumayilov 1986).

Even before the construction of the Boguchanskoye reservoir in the lower reaches of the Angara River, progressive long-term pollution and transformation of the river due to timber rafting (which had attained levels of 3,000,000 m^3 of wood annually) had already led to the loss of fisheries significance in Okynyovskaya and Angarskaya fishing sites, and of the Irikneyeva and Chuna rivers, which were former spawning grounds for acipenserids and salmonids. The Nevonskaya, Boguchanskskaya, Pinchugskaya, and Irikneeevskaya overwintering grounds of the sturgeon were destroyed and the mouth of the Taseyeva River was silted over. When the construction of dams in the upper Angara was completed, large numbers of sturgeon began entering the waters of the Taseyeva River and it's tributaries, the Chuna River and Biryusa River, where sturgeon had never been observed before (Burlaeva 1973). The Bratskoye and Ust-Ilimskoye reservoirs remain highly polluted by the forestry complexes (timber industry), and the concentrations of methyl-merkapthane and hydrogen sulfide attain hundreds of MCL (*GRCE* 1992).

The level of pollution in the Yenisei River is rather high. The section inhabited by Siberian sturgeon, i.e. below Krasnoyarsk, the water is continuously polluted by runoff from chemical, timber and metallurgical industries containing such specific chemicals such as lignosulphates, volatile acids, tars and asphaltenes. Their concentrations in 1989 were 1.3, 2.27 and 0.18 mg/L respectively (*ARSWQ* 1990a). This type of pollution continued at the same level in subsequent years (*GRCE* 1992). The river is seriously polluted by petrochemicals, the concentrations of which at the mouth of the Angara are in the order of 7 MCL, and around the Igarka and Dudinka, up to 8-11 and 2-3 MCL, respectively. In this same year, the level of phenol pollution in the Yenisei River (at the mouth of the Angara) exceeded the MCL by a factor of 7. At the Yenisei Bay in 1991, the concentration of phenols was 22 MCL (*GRCE* 1992). The tributaries of the Yenisei River, the Nizhnyaya Tunguska River and Podkamennaya Tunguska River, are characterized by copper complex concentrations of 9-14 MCL (*ARSWQ* 1990a). There are elevated levels of radioactivity related to the dumping of cooling waters from the uniflow reactor of the Krasnoyarsk Mountain-Chemical Plant, with high concentrations of registered short-lived radioactive isotopes of Sodium (Na_{24}), Chromium (Cr_{51}), Manganese (Mg_{54}), and Arsenic (As_{56}). The concentration of Na_{24} in 1991 exceeded the accepted norms by a factor of 9 (*GRCE* 1992).

The quality of the waters of the Yenisei River within the contemporary habitat range of the Siberian sturgeon, i.e. below Krasnoyarsk, is directly influenced by pollution in the upper reaches. In 1990, the upper reaches of the Yenisei below Kyzyl city were marked by an increase in phenol concentrations by 2 or 3 times of previous levels. The mean concentration of copper complexes increased to 9-14 MCL, with a maximum of 155 MCL. In the Sayano-Shushinskoye reservoir the content of petrochemicals peaked at 25 MCL. In the Krasnoyarskoye reservoir in 1990, the mean concentration of petrochemicals doubled from 1989 up to 8-13 MCL. In separate localities (near Ust-Abakan settlement), the concentration of petrochemicals attains 120 MCL (*ARSWQ* 1991). In recent years, the accumulation of chlororganic pesticides in fish, including DDT and its derivatives, has dropped in the Krasnoyarskoye reservoir (Chuprov and Vyshegorodtsev 1997).

On the whole, pollution levels have increased throughout the Yenisei River. In the middle and lower reaches, there has been an increase of phenols and petrochemicals up to 9-11 and 6-12 MCL respectively (Podtyosovo and Igarka). In 1990, a distinct increase in the number and biomass of midges and oligochaetes in the section of the river from Divnogorsk to Krasnoyarsk was observed. In 1990, there was an increase in pollution below Krasnoyarsk according to all indicators. Pollution in the Yenisei River is accompanied by a change in the contents and biomass of the benthos in separate sections of the river (*ARSWQ* 1991).

Khatanga River. The Khatanga is a relatively small river, the area of its basin is 346,100 km^2 and the length of it's channel it 779 km. It originates at the confluence of the Moyero and Kotuy rivers. After the Khatanga meets the Kheta, its width expands to 1 km. The bottom of the river is sandy. It is fed primarily by snowmelt; its mean flow is 3200 m^3/sec. The river freezes from the end of September to the middle of October. The break-up of the ice occurs between the middle of May and the middle of June (Davydov 1955).

Lena River. The largest river of Yakutia and second river in Siberia in water volume only to the Yenisei River, the Lena's basin covers 2,425,000 km^2 and its length is 4,270 km (Davydov 1955). The Lena River is fed primarily by snowmelt. In the upper reaches, snowmelt provides 40% of the volume, rain 35%, and groundwater 25%. Lower in the river, the proportions change because of the influence of the various tributaries, which have alternative water sources. For example, the Vitim River is fed to 88% by rain (Davydov 1955). Between the Yenisei and Lena basins there is a mountain massif demarking a "wind-divide" that strongly weakens the influence of the Atlantic air masses to the east (Antonov 1956). The climate in the Lena basin is, therefore, very severe and strongly continental. Annual temperatures have a negative mean and their amplitude over the year reaches 100° C. In the winter, there is a characteristic influence of arctic air masses with the establishment of long-term anticyclones and low temperatures. The permafrost is everywhere distributed in the Lena river basin (Doronina 1956). Therefore intense flow of melt-water together with summer precipitation causes a violent rise of the waters. The water level of the Lena undergoes significant oscillations. The water level amplitude is 4.8 m in the upper reaches at Kachug, 10.3 m 35 km above the mouth of the Kirenga River at the Markovo settlement, 7.4 m at Kirensk, 11.3 m at the mouth of the Vitim River, 14.3 m in the middle reaches at village Mukhtuyi, 10.7 m at village Nuya, 10.3 m at Pokrovskoye village, and 8.5 m in the section between the mouth of the Aldan River and the mouth of Viluy River (Davydov 1955). In the lower reaches of the river, the amplitude of the annual water level oscillation attains a maximum of 22.4 m (Vershinin 1964), and at Kyusyur settlement, 320 km from the mouth, 30 m (Mikhailov 1997).

The freezing of the lower Lena occurs during the last third of October. As in other major rivers flowing from south to the north and carrying an enormous flow of heat to the north, the Lena River freezes later than its tributaries. The Lena freezes, on average, 10 days after its tributaries. The break-up of the ice in the upper reaches occurs in the first third of May, and in the lower reaches, a month later in the first third of June (Davydov 1955).

The thermal regime in various sections of the Lena River differs significantly (Tab. 17). In the upper and middle reaches (settlements Kirensk, Olekminsk, and Pokrovskoye), the maximum average monthly temperature of the water reaches 19°C, from November to April they are close to 0°. In the lower Lena (Zhigansk and Kyusyur settlements) the mean monthly temperature is significantly lower and does not exceed 17°C. The freezing temperature period is longer, lasting from October to May inclusively.

The minimal mean monthly temperature is observed in the upper reaches of the Lena delta at the Tit-Ary settlement (Tab. 18). Mean annual sums of heat in the southern and northern sections of the Lena basin calculated using the data in Tables 16 and 17 differ by a factor of almost 2, comprising 1,973 and 1,073 degree days for the warmer (Olekminsk settlement) and colder (Tit Ary settlement) sections respectively. Duration of open water period is closely related with temperature regime (Tab. 19). The duration of the spring freshet as it advances from the upper reaches to the lower reaches of the Lena oscillates between 27 to 44 days, with a mean of 35 days (Antonov 1956). The rise of the water level at the time of the ice break-up as a consequence of ice-jams is significant and has a mean in the upper and middle reaches (mouth of the Vitim River) of 4.5 m, and in the lower reaches at Kyusyur settlement of 18 m (Antonov 1960).

The chemistry of the Lena is distinctive. The mineralization of the water is low and varies between 60 to 80 mg/L during high water and 580-680 mg/L at the end of winter. The waters of the Lena go from a hydrocarbonate class during the spring high water period to a chloridic class with the transition in the winter to groundwater sources with elevated quantities of anions, hard acids and calcium ions (Davydov 1955).

The largest tributaries of the Lena are the Vitim, Olekma, and especially the Aldan and Viluyi. These last two deserve a separate discussion, as sturgeons are distributed rather widely in these rivers.

Aldan River. The Aldan River, with a watershed area of 701,800 km^2 and a river length of 2,242 km, enters the Lena in the middle reaches (rkm 1,379). The river is fed in the upper reaches to 44% by rainwater while snowmelt and groundwater account for 35% and 21%, respectively. In the lower reaches, it is primarily fed by snowmelt. The river freezes at the end of October and opens up around May 17-21. Water levels are characterized by a spring freshet, no summer drought, and a sharp decrease in flow before the river freezes in October. The high water level peaks in June. In contrast to the Lena, the height of the high-water drops as one goes from the upper to the lower reaches. The water is weakly mineralized and following accepted classification they belongs to hydrocarbonate class (Davydov 1955, Alekin 1970). The river bottom consists of gravel and sandy-gravel along the entire length. The annual temperature regime is similar to that of the middle Lena River (Tab. 20).

The temperature regime is rather consistent across the river (Tab. 20). Differences between the upper (Uchur settlement) and lower (Batamai settlement) reaches are not great. The mean annual heat in the lower Aldan (Krest-Khaldzhai settlement) is 1,752 degree-days, only slightly lower than in the warmer parts of the Lena River.

Viluyi River. The Viluyi River enters the Lena at rkm 1,233. Its watershed area is about 490,000 km^2 and its length 2,435 km with a mean flow of 2,300 m^3/sec. The upper reaches are mountainous with the exception of Viluyiskoye reservoir (rkm 1,300 km reaching from above Chernyshevsky city, built in 1965-1976) up to the Suntar settlement (km 746). The river is fed to 54% by snowmelt. Rain accounts for 38% and groundwater for a mere 8% of the annual volume. The water levels show a dramatic spring freshet, raising up to 12.5 m in late May, gradually declining until early July. The river freezes during the last third of October and opens up in the second half of May. Around Suntar settlement, the Viluy generally freezes through to the bottom. The water is weakly mineralized and belongs to hydrocarbonate class (Davydov 1955, Alekin 1970). The annual temperature regime of the Viluy is similar to that of the middle Lena River.

Yana River. The Yana River is formed by the confluence of the Dulgalakh and Sartang rivers on the northern slope of the Verkhoyansk ridge. The area of its basin is 244,700 km^2 and its length is 879 km. The river is fed primarily by snowmelt and glacial runoff. No more than 10% of the annual flow is attributed to groundwater.

The spring freshet peaks at the end of May. The time to freeze over is extended and often lasts 3-4 weeks, beginning in early October in the lower reaches and in the latter half of October in the upper reaches. The Yana opens up in the second half of May in the middle and upper reaches and in June in the lower reaches (Davydov 1955). The length of the open period in the lower Yana (near Kazachye settlement) is, on average, 128 days (Averina et al. 1962).

Indigirka River. The Indigirka River is formed by the confluence of the Khastakh and Tarynyurakh rivers. The area of its basin is 360,400 km^2 and its length 1,790 km (Davydov 1955). The upper reaches, extending 940 km from the sources to the settlement of Krest-Mayor (rkm 850), are mountainous. The middle reaches, extending from Krest-Mayor to the Yakutskiye Yurti village, are in lowlands, full of lakes and swamps. In the lower reaches, the river divides into many branches which are 100-400 m wide. The lower Indigirka is characterized by extreme erosion of the shores due to the melting of the permafrost. The highly unstable shoreline recedes by as much as 70-100 m in the open period (Averina et al. 1962). During the rising of the river, the water carries an enormous quantity of plant debris.

The river is fed primarily by seasonal and year-round snows and glaciers. An important factor of the Indigirka's flow is the massive "naled" (additional layers of ice formed by water flowing at the surface of existing ice and freezing over again, layer by layer) occurring from the freeze-up of its tributary, the Moma River. The contribution of rain is significant while groundwater input is insignificant. The Indigirka is characterized by an extended high-water period, beginning in the middle of May and peaking in late July-August. The river freezes in the first half of October, while the retreat of the ice is slow. In the lower reaches at the Russkoe Ustye village, the river does not open up until mid June (Davydov 1955). The mean length of the open period in the Indigirka at the settlements of Vorontzovo, Chokurdakh and Russkoe Ustye is 132, 125, 121 days respectively (Averina et al. 1962). The climate in the Indigirka basin is extremely severe. The coldest area "pole of cold" in the Northern Hemisphere is located here in the Oyimyakon region. Table 21 presents the mean monthly water temperatures in the Indigirka from 1979-1988 according to the meteorological stations at Indirigirskaya and Chokurdakh.

The waters of the Indigirka are cold along its entire length (Tab. 21). In the south, at the Indirigirskaya meteorological station (rkm 1,412),

Table 21:
Mean monthly temperatures (°C) in the Indigirka River. Data represent averages over the period 1979 to 1988 (data from the weather stations mentioned, archives of Yakutian branch of the Eastern Siberia Fisheries Research and Designing Institute).

Weather station	Month					Net Annual Heat (degree days)
	June	July	August	September	October	
Indigirkskaya	4.9	13.0	10.7	5.0		1024
Chokudrakh	5.6	13.7	11.6	5.8	0.2	1126

and in the north, at the Chokurdakh meteorological station (rkm 153), the water temperatures are similar to those of the Lena River delta. The mean annual heat at both of these stations is similar and comparable to the heat flux in colder sections of the Lena River lower reaches. The colder temperatures at the southern end of the river are associated with the glacial melt-off that feeds the river. In the north, at the lower reaches of the river, the temperatures are only slightly warmer.

Kolyma River. The Kolyma River is formed by the confluence of the Kulu and Ayan-Yuryakh rivers on the Kolymsk and Chersk ridges. The area of its basin is 644,100 km^2 and its length is 2,600 km (Davydov 1955).

The river is fed primarily by snowmelt (47%), followed by rain (42%) and groundwater (11%). The spring freshet begins in May and is characterized by a rapid rise of water level and periodic ice-jams. The high water peaks occur in June during ice drifting (Davydov 1955).

Autumn ice starts drifting before freezing begins in late September in the lower reaches and in the early October in the upper reaches (Davydov 1955). A unique feature of the Kolyma River is the winter naled (layering of ice) and the creation of crevasses in the ice as a consequence of backwaters created by ice jamming. The break-up of the river ice is violent, beginning in the upper reaches in the middle of May and in the lower reaches in the beginning of June. The period of open water is brief and has a mean duration of 143, 139 and 121 days in the settlements of Zyryanka (995 km from the mouth), Srednekolymsk (650 km from the mouth) and Nizhnekolymsk (168 km from the mouth), respectively (Averina 1962, Shostakovich 1909).

Table 22 presents mean monthly water temperatures between 1979 and 1988 according to data from the meteorological stations at Cherskiy and Zyryanka. It can be seen that the water temperatures and mean annual heat in the lower Kolyma River – the principal habitat of the Siberian sturgeon in this river – is somewhat higher than in corresponding latitudes for the lower reaches of the Indigirka River and in the delta of the Lena River. The last is the principle foraging grounds of the lower Lena River population.

Changes in the environmental conditions for all fish species but particularly for the Siberian sturgeon in the rivers of Yakutia are primarily due to pollution. Hydroelectric construction in this region is poorly developed. The Kolymskaya hydroelectric station, built in the upper reaches of the river in 1981-1984, is far to the south of the sturgeon habitat range and has apparently not significantly affected the sturgeon.

Table 22:
Mean monthly temperatures (°C) in the Kolyma River. Data represent averages over the period 1979 to 1988 (data from the weather stations mentioned, archives of Yakutian branch of the Eastern Siberia Fisheries Research and Designing Institute).

Location	Month						Net Annual Heat (degree days)
	May	June	July	August	September	October	
Cherskiy		6.6	14.3	10.9	4.7	0.2	1128
Zyranka	0.8	9.4	14.2	11.9	5.3	0.4	1302

Viluyskoye Reservoir. The Viluyskoye reservoir, built in the upper Viluy between 1965 and 1976, is 480 km long, has a surface area of 217,000 ha, a maximum depth of 80 m, and a maximum width of 15 km. It is located in a permafrost zone. The Viluy dam has segregated the Viluy sturgeon population. Where it was once most abundant in the mouths of the Viluyi tributaries Chona, Chirkuo, and Akhtaranda, the populations moved up the river after the reservoir was filled and, in 1971, were located primarily in the variable backwater zone. Both mature and immature sturgeons were found here, including females with the remnants of unspawned eggs which indicated spawning further upstream from the reservoir (Biology of the Viluyiskoye Reservoir 1979). In the reservoir itself, the sturgeon populations were fished out within the first decade of the reservoir's existence and are no longer found. In the backwater parts of the river, they are now very rare (Kirillov 1986).

The levels and sources of pollution in the Yakutian rivers have been far less studied than in the Ob-Irtysh basin. The main sources of pollution in the Lena, Indigirka and Kolyma rivers are diamond and gold mining operations, communal discharges, and pollution arising from water transport. It has been proposed that elevated levels of copper and zinc concentrations have natural causes related to the dissolution of the soils and river bottoms that contain these metals. However, it has been observed that mining operations are also accompanied by elevations in copper and zinc concentrations in the sediments of adjacent water bodies (*ARSWQ* 1990a,b).

In Tables 23-25, we use data from the following publications: ARSWQ (1985, 1986, 1990a, b), *Hydrochemical Bulletin* (1977, 1978, 1979, 1983a, b, c).

It can be seen in these tables that the levels of pollution in the Lena, Aldan, and Indigirka are not as high as in the rivers of the Ob-Irtysh basin, though there is a marked upward tendency beginning in 1976. In the Ob basin, petrochemicals and phenol pollution levels are much higher, though the concentrations of zinc complexes are greater in the Yakutian rivers. The Kolyma around the settlement of Ust-Srednikan is marked by the presence of manganese and lead (*GRCE* 1992) and rather high concentrations of copper complexes. In 1989, around the settlement of Ust-Srednikan, as a result of a flushing of copper-containing pesticides and runoff from soils with copper complexes during freshet, their average annual concentration exceeded the MCL by a factor of 10. The maximum reported concentration was about 28 times higher than the MCL (*ARSWQ* 1990b).

As a whole, the ecological situation in the upper reaches of the Ob, the upper and middle reaches of the Irtysh, the upper reaches of the Yenisei, the Angara, the upper Lena, Viluyi, and Kolyma can be considered seriously impaired. In the Baikal region and in the middle Ob, the environmental disturbances can be considered as very serious or critical (*GRCE* 1992).

Table 23:
Maximum tolerance limits (MTLs) of pollutants in the Lena River (maximum values are in parentheses) (Annual Report on Surface Water Quality in the USSR [ARSWQ] 1985, 1986, 1990a, b; Hydrochemical Bulletin 1977, 1978, 1979, 1983a, b, c).

Year	Petrochemicals	Phenols	Zinc complexes	Copper complexes	Synthetic Surfactants
1976	2	-	-	-	-
1977	2 (30)	2.3 (31)	1	1	1 (5)
1978	5 (84)	2.2 (16)	1	1	1 (5)
1979	1-6 (39.6)	-	(19-24)	1-3 (39)	-
1982	5	5	-	-	-
1983	4 (62-117)	4 (53)	1 (2)	3	-
1984	3.5	3.5 (20)	1-2	2	-
1985	2.4 (58)	4 (35)	-	2.7	-
1990	1-8 (28)	1-4	1-7	2.1-7 (30)	-
1991	1-4	6-7 (24)	0-1	-	-

Table 24:
Maximum tolerance limits (MTLs) of pollutants in the Aldan River (maximum values are given in parentheses). (Annual Report on Surface Water Quality in the USSR [ARSWQ] 1985, 1986, 1990a, b; Hydrochemical Bulletin 1977, 1978, 1979, 1983a, b, c).

Year	Petrochemicals	Phenols	Zinc complexes	Copper complexes
1978	1 (96)	-	-	-
1979	(53)	-	(6.8)	(167)
1982	-	-	1.48	2.9
1983	5 (30)	4 (24)	-	1.5
1984	4 (30)	4	-	2
1985	2 (40)	5 (37)	-	2

Table 25:
Maximum tolerance limits (MTLs) of pollutants in the Indigirka River (maximum values are in parentheses) (Annual Report on Surface Water Quality in the USSR [ARSWQ] 1985, 1986, 1990a, b; Hydrochemical Bulletin 1977, 1978, 1979, 1983a, b, c).

Year	Petrochemicals	Phenols	Zinc complexes	Copper complexes
1976	-	-	-	-
1977	-	5 (10)	-	-
1978	-	-	2	-
1979	10	-	4 (23)	1-3 (149)
1982	-	-	-	-
1983	4 (12)	3	-	-
1984	3 (13)	5	-	-
1985	2 (46)	5 (39)	-	-
1990	1	3-6	1	2-7
1991	1-2	2-6	1	2-4

3. Diet and growth

3.1 Diet and diet composition

To date, many data have been collected on seasonal diet patterns in many parts of the species range (Isachenko 1912, Romanova 1948, Saldau 1948, Podlesniy 1955, Pirozhnikov 1955, Yegorov 1961, Kirillov 1962, 1972, Novikov 1966, Sokolov 1966b, Sokolov et al. 1986). In this chapter we summarize the historic findings by river basins and present also our own more recent data.

Ob-Irtysh basin. Yudanov's data (cited by Saldau 1948) from the Ob Bay at Bitkov Cape indicates the sturgeon feeds on the molluscs *Sphaerium corneum* and *Pisidium amnicum* and on the larvae of the tendipedids *Monodiamesa bathyphilum*, *Procladius* sp., *Chironomus* sp., and *Paracladopelma* sp. Golovkov (1942, cited by Saldau 1948), also reports feeding on *Sphaerium corneum* and *Pisidum amnicum* near the mouth of the Anti-Payuta River.

In the lower Ob, the diet of sturgeon is during summer dominated by *Sphaerium* sp. molluscs and by *Simuliidae* larvae during winter (Urban 1938, cited by Salday 1948). Sturgeon ranging between 11-84 cm TL, the main diet components are tendipedid larvae, *Paracladopelma* sp., *Trichocladius* sp., and caddisfly larvae *Hydropsychidae*.

In the Ob at Byelogorye settlement and in the Irtysh (Cheremkhovsk sands), the sturgeon feed primarily on some 20 species of midge larvae, of which *Paracladopelma* sp. dominates in number.

One also commonly encounters *Procladius* sp., *Trichoclaudius* sp. and *Culicoides* sp. in the guts, but not in significant quantities. During winter, the diet is largely composed of *Paracladopelma* and *Monodiamesa bathyphila*, whereas quantities of *Simuliidae* and *Coleoptera* larvae as well as *Sphaerium* sp. molluscs are of no significance. During summer (June-July) the quantity of *Paracladopelma* sp. in the diet falls to 0.8% while *Simulium* sp. larvae compose 85.6% of the stomach content by number and 37.2% by weight. This is due to an explosion in their population s in these warm months. During summer, the rate of molluscs and caddisworms also increases in the diet (Saldau 1948).

Sturgeon diet during hibernation contains exclusively midge larvae in young sturgeon (0.5 to 1.8 kg wet weight); counts range from 8,400 to 62,400 individuals. The diet of larger sturgeon (7 to 10.4 kg, maturation stage II) was more variable with gammarids and caddisworms as main components, followed, by mayfly larvae, midge larvae, water beetles, molluscs (*Pisidum* sp.) and plant detritus (in declining order). Sexually mature individuals (maturation stages III and IV) had empty stomachs (Votinov 1963).

Early life stages of the Siberian sturgeon (2 to 600 g) feed in the upper Ob River mainly on larvae of caddisworms (*Trichoptera*), mayflies (*Ephemeroptera*) and midges larvae (*Chironomidae*) (Solomonskaya 1952). For the year 1993 the diet of fish of similar size, 75-590 g, contained higher proportions of midge's larvae and oligochaete worms while caddisworms and mayfly larvae were absent. This is caused by a shift in benthic fauna induced by pollution, fostering the opportunistic midge's larvae and oligochaetes (Solovov 1997).

In historic studies (first half last century) in the Irtysh River (Vagay region), diets in July-August consisted primarily of Chironomidae, Trichoptera, and Ephemeroptera larvae (Revnivykh, 1937). Caddisworms are represented primarily by Brachycentrus subnubilus, Hydrophysche ornatula (16.7%) and mayflies by Palingenia longicaudata. Other species such as Simuliidae, Copepoda and Ostracoda were rarely encountered. The diet also included fish: roach up to 9-25 cm (up to 5 per stomach), dace up to 17 cm, ruffed up to 8 cm and burbot (Lota lota) up to 12 cm. Partial piscivory begins at ages 3+ (Revnivykh, 1937).

These observations agree with data on the variability of sturgeon diet previously observed in the Tobolsk region (Chalikov 1930) and in the Uvatsk and Vagay regions (Menshikov 1936) where the main diet was also Ceratopogon sp., Chironomidus sp., Trichoptera, larvae of Ephemeridae, Perlidae, Agrion (Odonata), while molluscs (Sphaerium sp.), crustaceans (Cyclops sp., Daphnia sp.) and fish were less important.

From the above data we may conclude that sturgeons of different size and age in the Ob-Irtysh basin exhibit several differences in their diets. The juveniles (under 50 cm) feed primarily on larvae of midges, caddisworms, mayflies, and black flies (simuliids) and on several molluscs (Spaerium). Larger sturgeon feed less on mayflies, consume more molluscs, and occasionally feed on eggs, larvae and juveniles of other fish. Fish encountered in the diet include roach, dace, ruffe, minnows, burbot, lamprey, and pike (up to 25 cm) (Dryagin 1949).

Yenisei River basin. Similar to the Ob River basin there is a difference in sturgeon diets in the Yenisei river and the bay. In the Yenisei Bay, the food mainly comprises the isopod Mesidothea (sea cockroaches) (Isachenko 1912, Nikolskii 1971). In the Yenisei Inlet sea cockroaches, accounted for more than 90% of the diet while midges larvae are almost absent. There is a historically recorded case of an 18.5 kg sturgeon whose stomach contained 70 small lampreys (Podlesniy 1958).

In the lower reaches of the river, the bulk of the diet is composed of isopods and amphipods (Mesidothea, Pontoporeia, and Marenzellaria). There are a few midges larvae, but plant detritus occurs frequently (Podlesniy 1955). In the river, the sturgeon mostly fed on lamprey larvae, Pisidium molluscs, and on the larvae of Tabanidae, Chironomidae and Ephemeridae (Isachenko 1912). In the Yenisei River stretch between the mouth of the Kureyka River to the Golchikha settlement, the main food were midges larvae: Polypedilum, Procladius, Trichocladius, Culicoides and others. Ephemerid larvae were the next most frequent item in the diet (Romanova 1948). Reports on the sturgeon's winter diet are conflicting. Isachenko (1912) reports from near Golchikha that stomachs were filled with sea cockroaches while Podlesniy (1958) believed the sturgeon do not feed at all in the winter. In general, a direct expected relationship was observed between the stomach contents and the composition of the benthos (Romanova 1948).

In the Angara River, the principal diet component were midges larvae (25,000-32,000 per stomach), simuliids and mayfly larvae. Gammarids were less important. Even more rare are molluscs and occasionally juvenile lamprey (up to 70 per stomach). Diet composition clearly reflects habitat related benthic fauna composition:

rocky beds contain more caddisworms while silty beds contain more midges larvae (Dryagin 1949).

In Lake Baikal, the main foods of the sturgeon are gammarid shrimp, midges larvae, stonefly and mayfly larvae, as well as Baikal bullhead (Cottocomephoridae). Molluscs, oligochaete worms, flatworms, and caddisworms, fly, beetle, and gnat larvae were all commonly observed as well (Yegorov 1961). Some stomachs were filled almost exclusively with sponge. The Baikal sturgeon is marked by a distinct change in diet with age. Young sturgeon at the Selenga River near mouth section feed almost exclusively on gammarids, but not infrequent a young sturgeon's stomach will be filled with juvenile bullhead. Molluscs, worms and flatworms are more rarely observed. The diet of yearling and juvenile fish was dominated by smaller forms of midges larvae and occasional gammarid young, which have more tender shells (Yegorov 1961). Older fish eat midges larvae and gammarids equally, and in the rivers they consume mayfly and stonefly larvae, all in larger sizes. At age 3-4 sturgeons begin to eat young bullhead although midges larvae and gammarids still dominate the diet. By 5-6 years and older, gammarids become the principal component of diet and young fish are more commonly found, primarily small bullhead but also, more rarely, some cyprinids. Adult fish feed primarily on bullhead, but also on the young of other fish, including perch (Yegorov 1961).

In the rivers that feed Baikal in the spring and summer the main diet of the sturgeon is mayfly, midge, and stonefly larvae. The larvae of caddisflies, gnats, beetles, gadflies (Tabanidae) and other insects are of secondary importance (Yegorov 1961).

Fish wintering in Baikal (January-March) do not stop feeding (Yegorov 1961).

All of this information is more than 40 years old and not much is presently known on the change in sturgeon diet in these rivers.

The more-thoroughly and partly more recently studied diets of Siberian sturgeon in the rivers of eastern Siberia, the Lena, Indigirka and Kolyma, allow for an analysis not only of the spectrum of the diet but of the frequency of separate components of the diet (Sokolov 1966b, Ruban and Konoplya 1994).

Yakutian river basins. The diet of Siberian sturgeon in the Lena River has been studied in greater detail (at least in the past). It has been shown that the dependence of general diet patterns on the benthic fauna and section of the river is similar to that of other river basins. In the near mouth parts of the rivers, the sturgeon feeds primarily on amphipods (*Pontoporeia*) and isopods (*Mesidothea*), and in other parts of the river they feed primarily on midge larvae, including *Paracladopelma* sp., *Cryptochironomus* sp., and *Stictochironomus* sp. (Pirozhnikov 1955, Sokolov 1966b). The importance of the midge larvae declines as one advances up the river and the importance of the molluscs, and caddisworms, mayfly and stonefly larvae increases (Sokolov 1966b). The proportion of various *Chironomidae* larvae is unequal and changes across different sections of the river, apparently reflecting a direct relationship to the composition of the benthos (Romanova 1948). Thus, according to the data of L. I. Sokolov (1966 b) in the delta of the Lena, the most significant are *Cryptochironomus camptolabis*, and less important are *Metriocnemus fuscipes*, *Procladius*,

Cryptochironomous, *Cryptochironomus rolli*, and *Chironomus heterodentatus*. Further upstream (in the section between the mouth of the Natara River to 200 km below Yakutsk, i.e. 600-1450 km from the mouth of the Lena) the diet of the sturgeon is predominantly *Procladius* followed by *Cryptochironomus camptolabis*. Of secondary importance are the diets of the Lena sturgeon which consist of the caddisfly larvae, *Hydropsyche* and *Arctopsyche*. Much more rare are *Leptocerus annulicornis* and *Phryganea striata*, the larvae of the stoneflies *Perla, Chloroperla,* and *Perloides*, and the larvae of the mayflies *Heptagenia, Ephemerella,* and *Ametropus* (Sokolov 1966b).

For the Lena sturgeon, as for the Ob and Baikal sturgeon, there is a characteristic increase in size of food with size of the sturgeon. This is particularly pronounced in the mouth of the Natara River – a region with a high benthic biomass. Here, as the sturgeon grow, the role of *Chironomidae* larvae drops sharply and molluscs become more important. At other river sections as sturgeons grow larger, they eat larger organisms within the same taxonomic group as well (Sokolov 1966b). The Lena sturgeon continues to feed throughout the winter (December). With rare exceptions, mature individuals with gonads in the IV and V stages of maturity do not feed in the period preceding spawning (Sokolov 1966b).

According to our data, in the Aldan River (the major right tributary of the Lena), sturgeon between 70 and 80 cm TL feed mainly on the larvae of Chironomidae, Plecoptera, Trichoptera, Heleidae, and Ceratopoginidae, with smaller quantities of *Sphaerium* and *Pisidium* molluscs. At a length of about 75 cm the sturgeon begins to supplement its diet with fish. Larger, adult fish over 90 cm become entirely piscivorous. One stomach contained 5-7 dace and roach, whose lengths ranged from 2.8-16.0% of the sturgeon's length (Sokolov et al. 1986).

A comparison of the diet spectrum and the frequencies of separate components in the diet of Siberian sturgeon from the lower Lena and Kolyma rivers (Tab. 26) shows that the two populations share 11 of 23 total possible foods and that the dominant food for both are the larvae of *Chironomidae* and *Heleidae*. However, there are significant differences in frequency. Thus, Copepoda, *Lynceus brachiurus* and fish are frequently encountered in the guts of the Kolyma sturgeon, whereas such foods as *Plecoptera*, *Diptera*, molluscs and oligochaetes, are common in the diet of the Lena sturgeon, but rare or entirely absent in the Kolyma.

The Kolyma sturgeon has a higher piscivory than the Lena sturgeon (Sokolov and Novikov 1965).

An analysis of our own samples of the Siberian sturgeon from different sections of the Indigirka River (Tab. 27) indicates that the most common components in the diets of the two samples studied (150 and 290 km from the mouth) are midge larvae (*Paracladopelma camptolabis*, *Prodiamesa gr. bathyphila,* and *Procladius*), and larvae from *Ceratopogonidae, Plecoptera* and *Ephemeroptera*. On the whole, the composition of the diets in the samples collected 150 km from the mouth of the river was impowerished in comparison with the samples collected higher upstream, where besides the above noted midge

Table 26:
Frequency of occurrence of food species in the stomachs of Siberian sturgeon from the Lena and Kolyma rivers (modified after Sokolov and Novikov 1965). n = number of specimens investigated and size range in cm.

Food	Frequency of occurrence (%)	
	Lena sturgeon n=244, TL from 25 to 122 cm	Kolyma sturgeon n= 17 TL from 36 to 92 cm
larvae	97.5	76.9
chrysalids	25.4	-
Ephemeroptera- larvae	7.0	-
Trichoptera- larvae	11.1	17.6
Plecoptera- larvae	13.5	5.9
Megaloptera- larvae	0.4	-
Heleidae- larvae	45.1	33.3
Diptera- larvae	25.0	5.9
Coleoptera	0.8	5.9
Copleoptera imago	0.4	-
Oligochaetae	17.0	-
Formicidae	2.9	5.9
Mollusca	16.4	-
Hydracarina	1.6	5.9
Mesidotheo entomon	-	5.9
Copepoda	2.5	17.6
Cladocera	0.4	-
Lynceus brachiurus	-	53
Ostracoda	4.5	-
Gammaridae	0.8	-
lamprey larvae	1.6	-
Pisces sp.	1.2	17.6
fish eggs	4.1	5.9

larvae, the diet also included *Cryptochironomus gr. defectus*, *Chironomus macani*, *Parachironomus pararostratus* and *Stictochironomus gr. Histo*.

The composition of the diet in the samples we analyzed from the Kolyma River differs along the length of the river (Tab. 27). In the sample from the section of the river adjacent to the mouth of the Ozhogina River (900 km from the mouth), the composition of the diet is more limited. *Chironomus*, *Liniella arenicola*, *Chicotopus gr. silvestris* and *gr. algarum*, *Plecoptera* and *Trichoptera* larvae were absent, as were the oligochaetes, nematodes and molluscs. On the other hand, all of the foods species were typical for the middle reaches of the river, the only species absent in the lower reaches were the larvae of *Cryptochironomus gr. defectus*. The maximal number of food components was observed in sturgeon from the lower reaches of the Kolyma River.

The high frequency of *Paracladopelma camptolabis* and *Prodiamesa gr. bathyphila* midge larvae in all four of the examined samples indicates that the sturgeon in the Indigirka and Kolyma forage along the channel part of the rivers on sandy-gravely substrates as much as on the silty bottoms of the branches and the bays. The frequency of these larvae agrees with the character of the river bottom in these locations and the benthic fauna they generally support. Thus, in the lower Indigirka where there are many river branches with silty bottoms, *Paracladopela camptolabis* is more frequently encountered than in the Kolyma River. The highest numbers of *Prodiamesa gr. bathyphila* are observed in the sturgeon of the Kolyma were in samples collected near the mouth of the Ozhogina (900 km from the mouth). The river there has a sandy-gravelly bottom and serves as a spawning ground for the sturgeon.

A visual analysis of the contents of the stomachs from sturgeon on their spawning grounds in the Indigirka and Kolyma rivers indicate that feeding occurs during the spawning period as well.

Black gnat larvae (*Ceratopogonidae*) play a significant role in the diet of the sturgeon in the Indigirka and Kolyma rivers. As one can see (Tab. 27) the frequency of these larvae in southern sections is higher. Stonefly larvae *Plecoptera* are more commonly found in Indigirka sturgeon than in Kolyma sturgeon. In the fast-running section of the Kolyma with a sandy-gravel bed (near the mouth of the Ozhogina), they are entirely absent.

Comparisons of our data on the Kolyma sturgeon diet with data collected by Novikov (1966) between the settlement of Zyryanka (995 km from the mouth) and the beginning of the delta, shows a high frequency of midge, black gnat and caddisfly larvae in the diet. The latter we only encountered in our samples from the lower reaches of the river. Unfortunately, Novikov does not indicate the specific locations of his collections. It is likely that the caddisfly larvae were characteristic of the lower reaches in his samples as well.

In sections of the lower Kolyma where the influence of marine fauna is significant (up to 60 km from the mouth of the river) we also encountered in their stomachs such organisms as *Mesidothea entomon*. This is corroborated by Novikov (1966). At the upper end of the delta, sturgeon forage at rather significant depths (26-32 m). Much of the marine benthos fauna found in the diet of sturgeon is typical for lower reaches and the brackish pre-delta waters of the respective rivers. Thus, in the Yenisei below Dudinka, the diet of the sturgeon was composed mainly of *Mesidothea*, *Pontoporeia*, and *Marenzellaria* (Podlesniy 1958). In the Yenisei Bay and in the river in the winter at the villages of Golchikha and Pustoye, the diet was dominated by *Chirodothea sibirica* as indicated by the historic samples of Isachenko (1912). In the Yenisei Inlet, sea cockroaches (Mesidothea) made up about 90% of the sturgeon's diet (Podlesniy 1958). At near-mouth areas of the Lena River (Tiksy Bay near the Mostakh cape) *Pontoporeia affinis* composed up to 80% of stomach content by weight and *Mesidothea entomoni* occasionally up to 50% (Pirozhnikov 1955).

Some of the qualitative differences between the composition of the diets of the fish we examined from the Indigirka and Kolyma can by attributed to differences in size, since, as was shown for the lower Lena sturgeon, with increasing size of fish the size of organisms in it's diet changes (Sokolov 1966b). In the given case, this dependence is particularly obvious with respect to the sizes of molluscs. However, we observed molluscs in the stomachs of only three specimens from the

Table 27:
Numbers of food items in the stomachs (%) of Siberian sturgeon from the Indigirka and Kolyma rivers (our data) sampled between 1984 and 1989. N = number of fish, TL = total length.

River:	Indigirka		Kolyma	
Distance from mouth:	157 km	285 km	235 km	900 km
N	8	12	13	8
TL, cm	21-49	34-62	37-124	49-72
Year of collection	1984-1985	1985, 1987	1989	1988
midge larvae:				
Polypedilum nubescolosum	-	16.7	15.4	12.5
Paracladopelma camptolabis	75.0	58.3	30.8	37.5
Cryptochironomous gr. defectus	-	16.7	-	12.5
Cryptochironomous gr.dneprinus	-	-	62.5	7.7
Chrinomus macani	-	16.7	-	-
Parachironomous pararostratus	-	8.3	23.1	12.5
Stictochironomus gr.histro	-	8.3	15.4	75.0
Prodiamese gr. bathyphila	62.5	66.7	53.8	87.5
Procladius sp.	50.0	66.7	23.1	25.0
Chironomus sp.	-	-	30.8	-
Lipiniella arenicola	-	-	30.8	-
Tanytarsus sp.	-	-	-	12.5
Cricotopus gr. silvestris	-	-	15.4	-
Cricoptopus gr. algarum	-	-	15.4	-
midge chrysalids:	12.5	16.7	23.1	-
other larvae:				
Cerapotogonidae	37.5	41.7	23.1	50.0
Plecoptera	50.0	83.3	15.4	-
Ephemeroptera	37.5	33.3	30.8	50.0
Trichoptera	-	-	23.1	-
Diptera:				
Diptera imago	12.5	33.3	-	12.5
Diptera larvae:				
Brachycera sp.	-	16.7	-	-
Hebecnema sp.	-	8.3	-	-
Insecta imago	-	-	-	12.5
Mollusca	-	-	15.4	-
Copepoda	-	-	-	25.0
Oligochaeta	12.5	8.3	15.4	-
Nematoda	-	8.3	7.7	-
fish eggs	12.5	8.3	-	-

Kolyma, which does not provide us with enough power to explore this dependence. For the remaining foods, a visual analysis was unable to identify increasing prey size with sturgeon growth in our study. The only obvious difference in the diet as a function of size was that in contrast to small sturgeon larger specimen consumed fish.

Out of 293 Indigirka sturgeon stomachs, 5.5% contained fish. Foraging on fish is more typical for sturgeon over 90 cm TL and 20 years of age. Their stomachs may contain as many as four dace as long as 17.5 cm. Occasionally, young fish occur in stomachs of smaller sturgeons. Thus, fish juvenile were found in the stomach of the sturgeon 38.8

cm in length and 5 years old. In the diet of the Indigirka sturgeon, exceptional items such as rodent were also found. This phenomenon was observed in 1987 during a season of severe forest fires that induced migrations of rodents across the river (Ruban and Konoplya 1994).

Despite the fact that the sturgeon is a typical benthophagic fish, foraging on fish is to some extent typical for all of the populations which have been studied: the Ob (Saldau 1948, Votinov 1958), the Lena (Sokolov 1966b, Kirillov 1972, Ruban and Akimova 1991) and it's tributaries, the Viluyi (Kirillov 1962), the Aldan (Sokolov et al. 1986), the Indigirka (Ruban and Akimova 1991) and the Kolyma (Novikov 1966, our data). Perhaps to the highest degree foraging on fish is typical in the Baikal population. In winter at Selenga shoals the sturgeon diet is dominated by bullhead (*Cottocomephoridae*), the weight of stomach content can achieve 1.5 kg. There is a recorded case of a 37-kg Baikal sturgeon containing 220 one-year-old perch (Yegorov 1961). Fish were not observed in the diet of the Yenisei sturgeon (Isachenko 1912, Romanova 1948, Podlesniy 1955).

These data support the conclusion that the sturgeon is to some extent a predatory fish, in contradiction to the opinion of Kirillov (1972), who thought that the sturgeon only ate traumatized or dead fish. Support against this hypothesis is given by our observations that fish are found in the stomachs of sturgeon far from fishing grounds, such as in the Aldan. Furthermore, where there are fishing grounds, the mesh in the nets is wide enough that the small fish consumed by the sturgeon cannot be traumatized.

Because of the difficulty of accessing much of the Siberian sturgeon's habitat and also because many populations are not sufficiently large enough to take warrant the collection of daily data, there have not yet been developed estimates of the diet rations available in the natural environment. However, one can use indices of stomach fullness (SFI) to estimate the intensity of feeding for individual populations (Tab. 28).

From Table 28 it is evident that SFI varies widely between populations. The greatest values are observed for Ob sturgeon and for Lena sturgeon near the mouth of the Natara River, and in the section of the Lena River 200 km downstream from Yakutsk. However, a high value for SFI among the Lena sturgeon does not necessarily indicate high availability of food, since the indigestible part of the stomach contents (silt, sand, plant detritus) accounts for a large proportion of the total, occasionally over 90% (Sokolov 1966b). The Baikal sturgeon also takes in a large amount of silt and sand (Yegorov 1961). It is likely that among mature individuals of Baikal sturgeon, whose diet consists primarily of fish (Yegorov 1961), the amount of sand and silt must be smaller. The stomach fullness index does not depend on the size of the fish (Yegorov 1961, Sokolov 1966b).

The above reported results indicate that our knowledge on the feed composition is fragmentary. Considering the huge area and the large river systems, this is quite understandable. There is a need for intensified studies particularly because of the changes in habitat and hydrodynamic regimes which – over the decades – have not only affected the migration and distribution of sturgeons but certainly also affected ecology of the benthos and thus the feed base for sturgeons. It is unfortunate that we have to base our conclusions on the rather sporadic and partly older studies (older than half a century). However, the information gathered can still serve as a comparative baseline with data obtained now or in the future.

3.2 Growth

Siberian sturgeon can attain rather significant body sizes and mass. Sturgeon caught from the Ob River have been reported to weigh as much as 180-200 kg (Berg 1949, Dryagin 1949). However, the species is in its natural distributional range, one of the slowest growing acipenserids. This is due primarily to the ecological and environmental conditions of the habitat, in particular the availability of food and the low temperature regimes of the water bodies it inhabits. Under artificial conditions with unlimited food availability and high temperatures the growth rate of the Siberian sturgeon can significantly increase. In rearing experiments with the Lena River sturgeon in aquaria it was demonstrated that its growth rate is comparable to that of the Russian sturgeon from the Volga River (Sokolov 1965a). In the Narva fish hatchery, the Siberian sturgeon grew significantly faster than the Russian sturgeon (Egelskiy 1967, 1970). In the pools of the Aksai fish hatchery in the Rostov region, four-year old Lena sturgeon had an average weight of 1,900 g, with some individuals attaining 3,000 g. Siberian sturgeon cultivated in cages in the Pyalovsk Reservoir grew faster than any of the other sturgeon species reared under the same conditions and four-year-olds attained a mean weight of 1,880 g (Mikheyev 1974, 1979). The growth potential of the Siberian sturgeon can be particularly well demonstrated when rearing at high temperatures (Berdichevskiy et al. 1983) (Tab. 29). An elevation of the temperature by 4-7°C in the summer and by 9-10°C in the winter in comparison to the water temperatures in the wild are accompanied by a 7-9 times increase in growth rates (Akimova et al. 1980). In warm-water culture of the Lena river sturgeon in France, high growth rates were also obtained, with a mean mass for 4 year olds of 2,500 g (ranging from 1,250 to 3,850 g) (Lamarque 1979). With similar culture methods, the sturgeon in the Tallinn (Estonia) fish hatchery yielded mean annual growth increments of 0.75 kg (Kokarev 1983).

Table 28:
Average stomach fullness index (SFI) in pptt = parts per ten thousand of Siberian sturgeon from various populations and according to various studies during the 1960s (in brackets SFI range).

Location	N	SFI	Source
Ob		73 (0.4 – 216.0)	Votinov (1963)
Ob Bay		134.7	Saldau (1948)
Lake Baikal		71 (12-242)	Yegorov (1961)
Lena River:			
Stolb Island	166	106.2	Sokolov (1966b)
674 km from mouth (at Natara River mouth)	67	158.2	Sokolov (1966b)
1233 km from mouth (at Viluyi River mouth)	2	44.6	Sokolov (1966b)
1440 km from mouth (200 km downstreams Yakutsk city)	9	128.1	Sokolov (1966b)

Table 29:
Mean and maximum weight (g) of Lena sturgeon at age in hatcheries using warm water (Berdichevskiy et al. 1983).

Age (years)	Konakovo fish hatchery (tanks) mass (g)		Tallin fish hatchery (cages) mass (g)	
	Mean	Maximum	Mean	Maximum
1	34	100	100	300
1+	400	1200	600	1200
2	1400	1600	1300	2000
2+	1500	3100	1900	3600
3	2500	-	2400	4100
3+	2700	4900	3100	5100
4	3100	5800	3800	6000
4+	3800	6700	4200	8000
5	4100	7600	-	-
5+	5500	9100	-	-
6	5600	-	-	-
6+	6400	10800	-	-

In natural water bodies, growth in length and in weight of Siberian sturgeon varies significantly between populations.

According to our results (Fig. 36), the slowest rate of body length growth is exhibited by Siberian sturgeon populations from the lower reaches of the Lena and Indigirka rivers. Yenisei River sturgeon exhibit significantly higher growth rates. Sturgeons from the Baikal and the Kolyma River populations are characterized by higher growth rates than fish from the Lena and Indigirka rivers. Young specimens of Baikal and Kolyma sturgeon up to 11 years of age, are of similar length, but after the 12th year, the Baikal sturgeon surpasses the Kolyma sturgeon, and after the 13th year, it surpasses the Yenisei sturgeon. Among river populations of Siberian sturgeon, the fastest growing population is the one from the Ob River, but the young, up to 8 years of age, are similar in size to the Yenisei population. At an age of 17, the Baikal sturgeon surpasses the Ob sturgeon as well. The Kolyma sturgeon is the fastest growing among those fish of the rivers of the Yakutia region. As one can see from Figure 36 the Kolyma sturgeon (at the age of 17 years) surpasses the Yenisei sturgeon and at age 21, the Ob sturgeon.

Differences in growth rate (length) of males and females are poorly expressed (Figs. 37, 38, 39, 40). As a rule, among young immature individuals, the females are of somewhat larger size. In older age classes, differences between the sexes diminish. But in separate populations, such as that of the lower Lena River, the males are of larger size (Koshelev et al. 1989).

Growth in weight of the Siberian sturgeon (Figs. 41, 42, 43, 44, 45) exhibits similar differences between populations and sexes.

The shape of growth curves (total length) in sturgeon from the lower Ob, Yenisei, Lena and Indigirka rivers show that growth rates start to decline at ages of 11-13, i.e. slope of the growth curve obtained from the observed data drops (Figs. 36-40). As noted from the rivers of Yakutia, the decline in growth rates is presumably associated

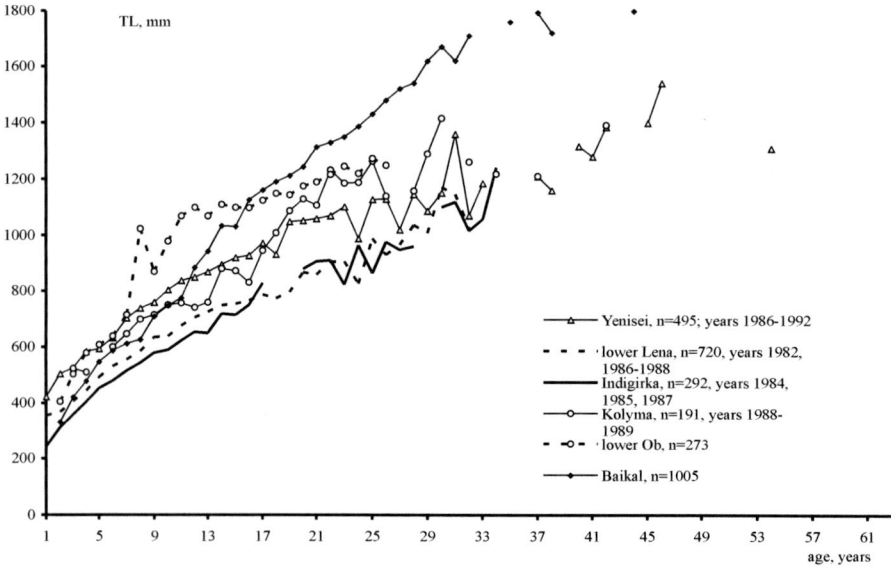

Fig. 36: Relation between total body length (TL, in millimetres) and age (years) of the Siberian sturgeon (sexes combined, immature juveniles included) in populations from river systems, respectively different environmental regimes. Data on lower Ob were taken from Dryagin 1949 and on the Lake Baikal from Yegorov, 1961.

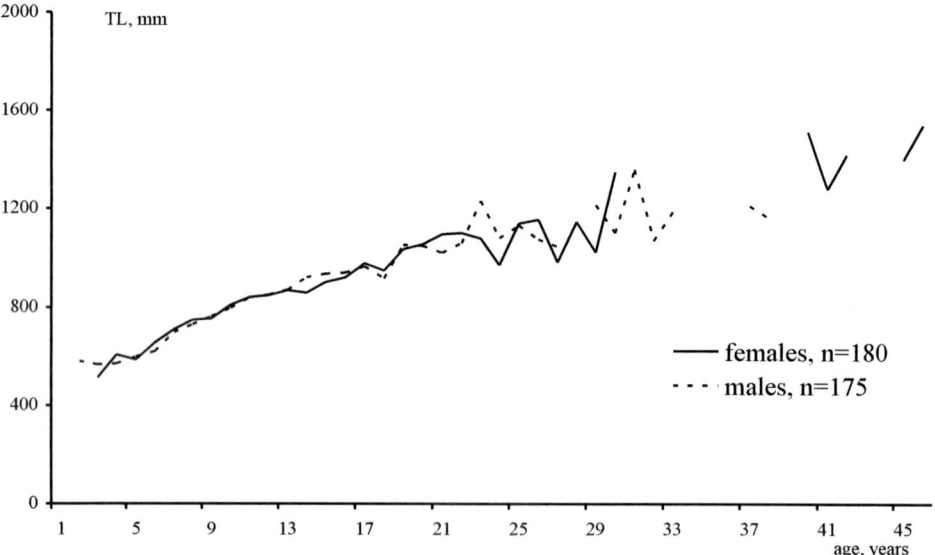

Fig. 37: Relation between total body length (TL, in millimetres) and age (years) of the Yenisei River Siberian sturgeon males and females. Data obtained between 1988 and 1992.

Part II - Ecological Characteristics of the Siberian Sturgeon

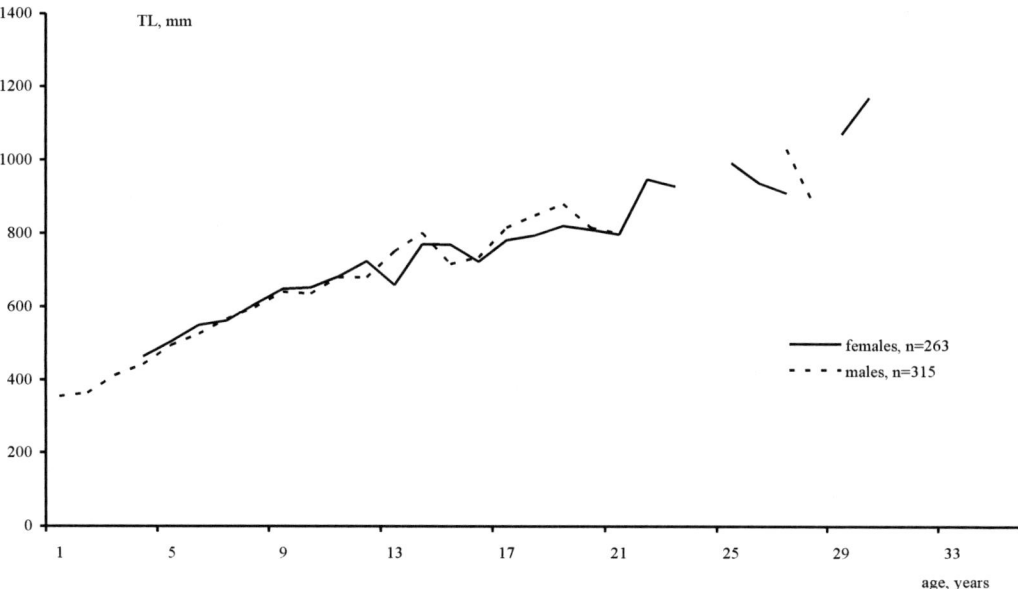

Fig. 38: Relation between total body length (TL, in millimetres) and age (years) of the lower Lena River Siberian sturgeon males and females. Data obtained between 1982 and 1988.

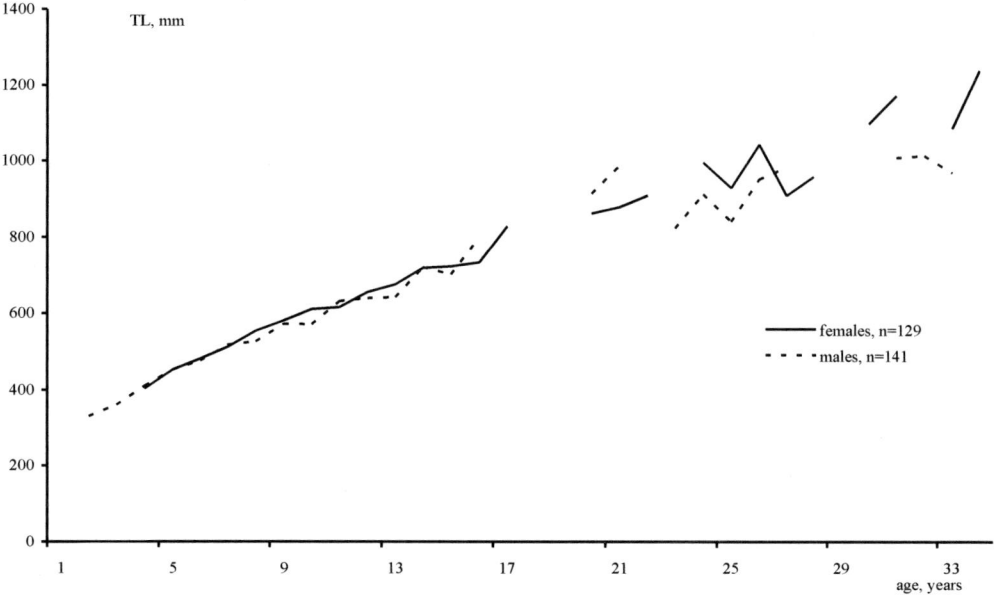

Fig. 39: Relation between total body length (TL, in millimetres) and age (years) of the Indigirka River Siberian sturgeon males and females. Data obtained in 1984, 1985 and 1987.

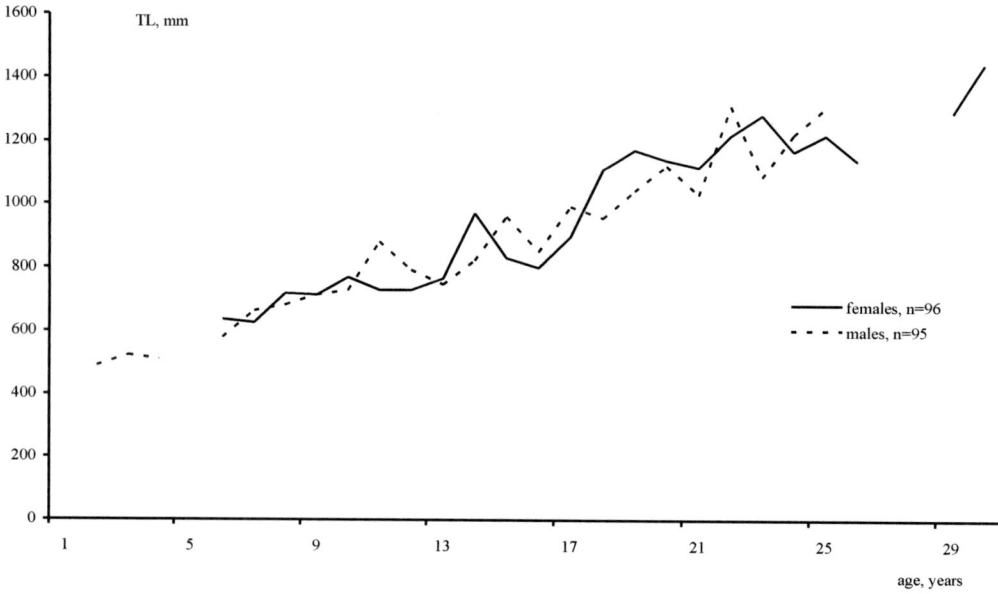

Fig. 40: Relation between total body length (TL, in millimetres) and age (years) of the Kolyma River Siberian sturgeon males and females. Data obtained in 1988 and 1989.

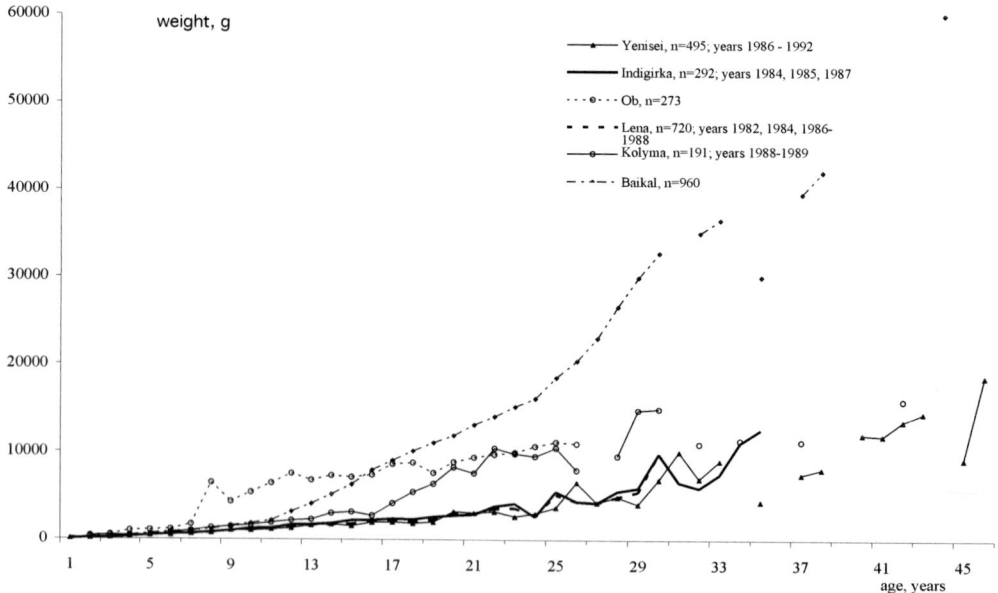

Fig. 41: Relation between wet weight (g) and age (years) of the Siberian sturgeon (sexes combined, immature juveniles included) in populations from river systems, respectively different environmental regimes. Data on lower Ob were taken from Dryagin, 1949, and on the Lake Baikal from Yegorov, 1961.

Part II - Ecological Characteristics of the Siberian Sturgeon

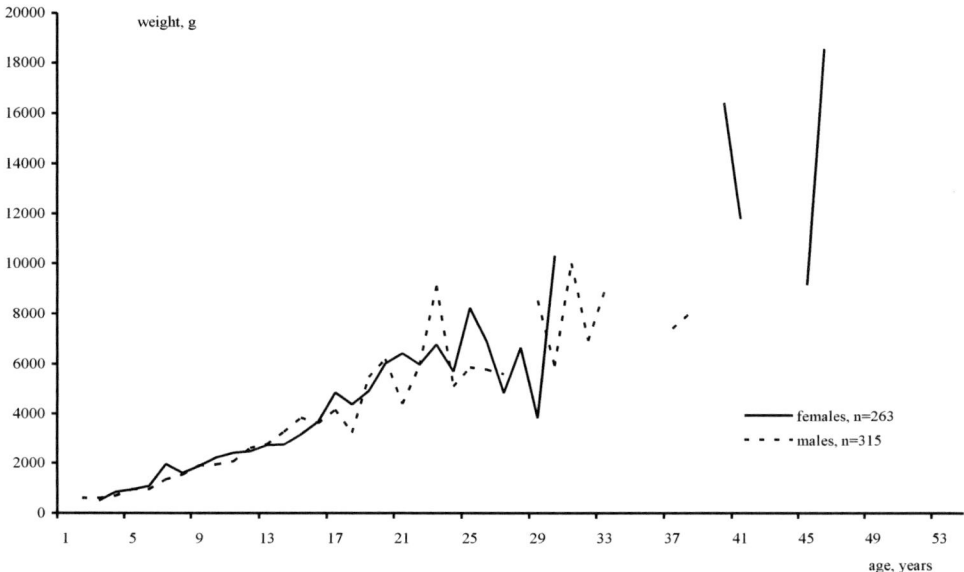

Fig. 42: Relation between wet weight (g) and age (years) of the Siberian sturgeon in the Yenisei River separately for males and females. Data obtained between 1986 and 1992.

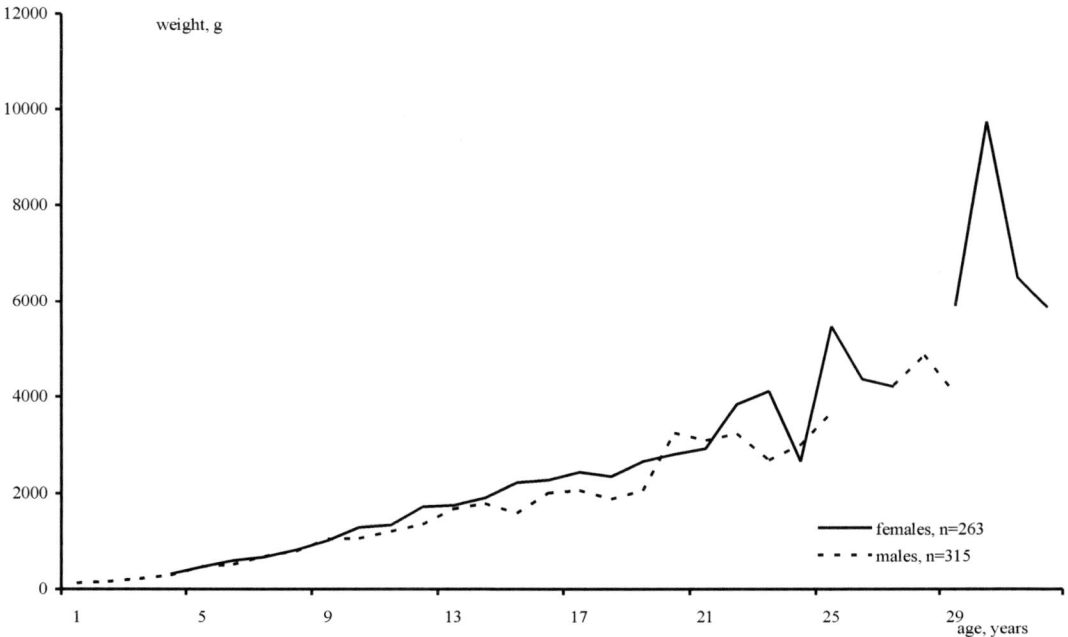

Fig. 43: Relation between wet weight (g) and age (years) of the Siberian sturgeon in the lower Lena River separately for males and females. Data obtained in 1982, 1986-1988.

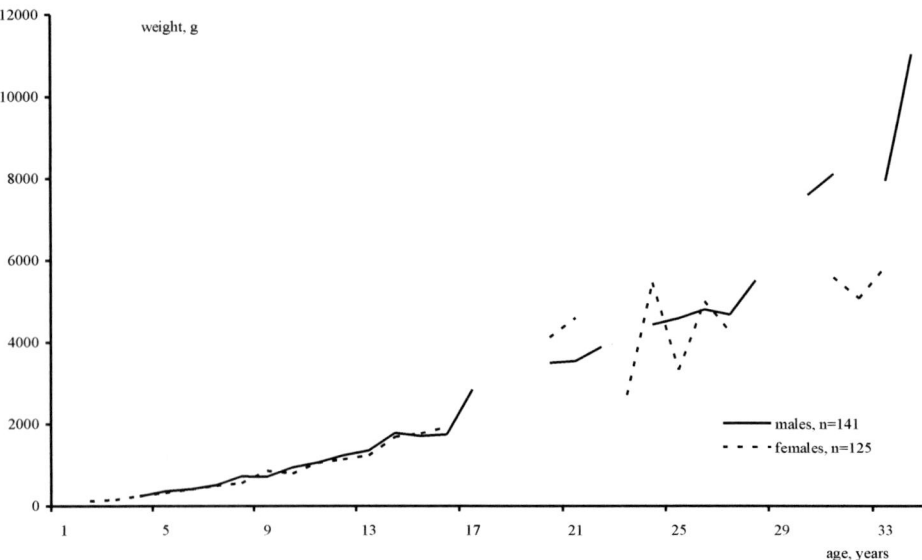

Fig. 44: Relation between wet weight (g) and age (years) of the Siberian sturgeon in the Indigirka River separately for males and females. Data obtained in 1984, 1985, 1987.

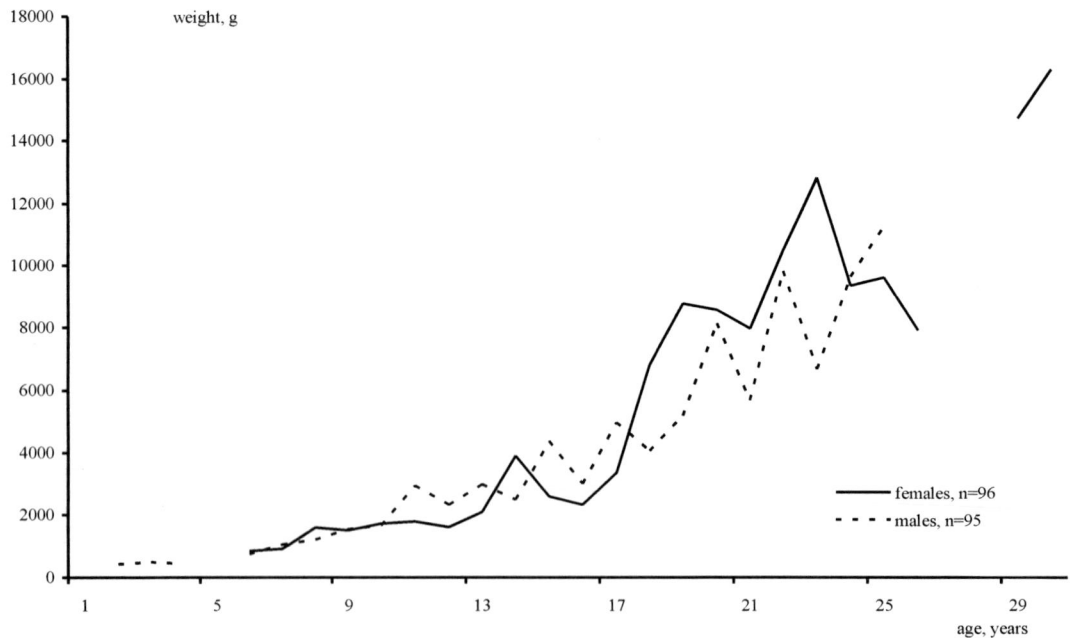

Fig. 45: Relation between wet weight (g) and age (years) of the Siberian sturgeon in the Kolyma River separately for males and females. Data obtained between 1988 and 1989.

with the onset of sexual maturity (Sokolov et al. 1986, Koshelev et al. 1989, Ruban and Akimova 1991). However, these phenomena can be a result of another causes, for example, a shorter life span of faster growing fish in separate age classes.

In order to test these hypotheses, we (Ruban 1995b) first established a positive correlation between the radius of sections of marginal ray of pectoral fin and total length, followed by a comparison of the distributions of annulus measurements for a particular young age group with an analogous distribution of the same annulus for an older age group. Similarities or differences in the distribution of the same annulus in younger and older age classes served as indicators of conservation or elimination of individuals with different growth rates. An analysis of a sample of sturgeon (257 individuals) from the lower reaches of the Lena River collected at Tit-Ary island and in the branches of the delta in 1986 show a difference in the growth rates of total length in males and females. At the age of 13 years the males surpassed females in growth (Fig. 46) (Koshelev et al. 1989). Females lived longer than males, attaining 51 years, whereas there were no males over 27 years. Consequently, the shape of the growth curves obtained for the entire sample can be affected by the unequal life span of females and males. The shorter life span of the latter, growing initially faster create of the false impression of slowing of the growth rate in the pooled growth plot. Besides that this effect also could be caused by shorter life span of faster growing females, examined separately, which can be tested with the same method described above.

There is a rather high correlation ($r = 0.63$) between the diameter of the marginal ray of the pectoral fin measured at 8-10 mm from the articulation (at the point of cross sections for age determination) and the total length. The total length and measured third annulus of three-year old males and females were very similar, and the samples used to determine this distribution were not separated by sex (Fig. 47 a). Analogous distributions were plotted separately for males (Fig. 47 b) and females (Fig. 47 c) that had attained sexual maturity (older than 10 years).

A comparison of the distributions shows that the mode of the diameter of the third annulus in older age groups is smaller than that of the three-year olds, and there are no individuals that attain the maximum diameter of the third annulus, i.e., individuals which grew faster when they were young. This is more clearly evident among males. The lack of slow-growing individuals with small annulus diameter in the sample of three-year-olds is most likely due to their small numbers in the sample (18 individuals). Thus, in assumption of sustaining a high growth rate of fast-growing individuals throughout their lifespans, it is possible to suppose that the visible flattening of the growth curve slope previously interpreted as a slowing of growth is in fact due to increased mortality among fast-growing fish. This applies equally to males and females.

Lowered life spans among fast-growing individuals has been observed for many species of fish (Nikolskii 1974). In the last case the visible flattening of the growth curve is redoubled by lower life span of males growing faster in older age groups. On the other hand, one cannot entirely exclude an actual slowing of growth with age.

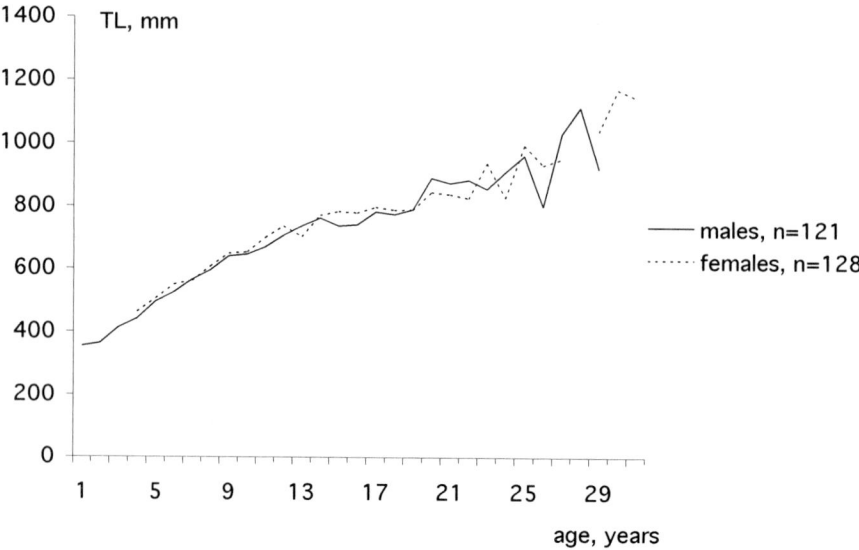

Fig. 46: Relation between total body length (TL, in millimetres) and age (years) of Siberian sturgeon males and females in the Lena River delta (TL – total body length in millimetres). Data obtained in 1986.

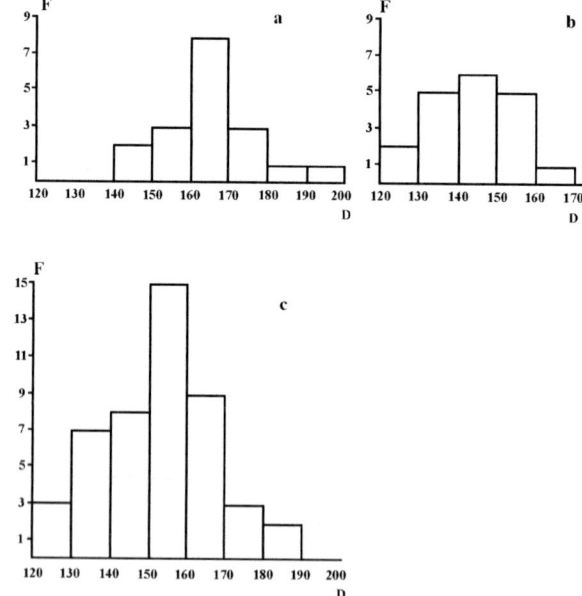

Fig. 47: Frequency distribution (F, examples) of the third annulus diameter (D, mm) of marginal ray of the pectoral fin in Siberian sturgeon from the Lena River delta; a – three year old specimen; b – males older than 10 years; c - females older than 10 years). Data obtained in 1986.

The life span of males and females is not equal in populations of Siberian sturgeon where growth rates differ. In the Ob, Yenisei, and Baikal populations, characterized by high growth rates, there was no significant difference in the life span of males and females (Votinov et al. 1975, Podlesniy 1955, Yegorov 1961). Our data on age composition of the Yenisei sturgeon catches corroborate these findings.

Table 30, shows that the maximum ages of males and females in our catches from different populations are significantly different. The greatest difference was observed in the populations of the Lena and the Indigirka rivers, characterized by slowest growth. In examining analogous differences in the population structure of stellate sturgeon, Nikolskii (1974) proposed that, as in the majority of sturgeon species, they were due to differences in food availability. Our data on growth rates of the sturgeons in different populations supports this hypothesis.

Food availability and the thermal regimes were included in the analysis of differences in growth rate and structure of separate populations of sturgeon, assuming direct influences on both growth and the development of food resources. Currently, as it was noted above, there are no data on the daily food availability for sturgeons from different populations. Consequently, a comparison of food availability can be accomplished only using the data on total benthic biomass – the primary source of food for sturgeons.

The main foraging grounds of the sturgeons in the Ob basin are found in the floodplains of the Lower Ob, its delta, and in the Ob Bay. In the lower Ob, the most important food source is the benthic biomass in the delta branches ranging from 123 to 2,165 kg/ha. The primary species in the benthos are small molluscs (*Sphaerium*, *Pisidium*) and tendipedid larvae. In the Ob delta, the greatest densities of benthos are found in the deeper sections of the delta branches, where the biomass attains, on average, 276 kg/ha, and the primary content of the benthos is molluscs (70% of the biomass). In the freshwater southern part of the Ob Bay, the biomass is 18.1 kg/ha. In the middle part of the Ob Bay (from Cape Kammeniy to Cape Khonorosale), the mean biomass is 23.7 kg/ha, dominated by mysids, amphipods *Pontoporea affinis* and *Mesidothea*. Brackish water benthic organisms dominate north of the 72nd parallel in the Ob Bay, and the biomass ranges from 6 to 632.6 kg/ha (Votinov 1963).

In the Novosibirsk reservoir, the benthic biomass is inconsistent and changes between years in dependence of flow. In the first years after the filling of the reservoir, the benthic biomass attained a maximum (640-850 kg/ha), as is also

Table 30:
Maximum age (years) of Siberian sturgeon males and females from different populations (our data).

River	Year of sampling	Males	Females
Yenisei	1988-1992	51	49
Lena	1986	28	50
Indigirka	1984, 1985, 1987	38	63
Kolyma	1988, 1989	30	42

typical for other reservoirs. Since its second decade of existence, there has been a relative stabilization of the benthos and it fluctuates between 107.4 and 220.1 kg/ha (Blagovidova 1976).

The benthic biomass of the Yenisei varies significantly in different sections of the river, each of which is dominated by different bioceonosis. As a whole basing on the proportions of the surface areas of various types of bottoms, it is estimated that the benthic biomass in the middle reaches of the Yenisei in its upper and lower sections is 15.7 and 9.9 kg/ha respectively and 18.5 kg/ha in the lower reaches of the river (Greze 1953). In the Yenisei delta and the Yenisei Bay, the benthic biomass is 56.8 and 46.9 kg/ha respectively. According to later measurements (Greze 1957), the benthic biomass of the Yenisei was 20.8 kg/ha in the lower end of the upper reaches between Krasnoyarsk city and the Angara River, 11.5 kg/ha in the upper end of the middle reaches between the mouths of the Angara and Podkamennaya Tunguska rivers, 10.6 kg/ha in the lower end of the middle reaches between the mouths of the Podkamennaya Tunguska and Nizhnyaya Tunguska rivers, and 13.5 kg/ha in the lower reaches of the river between the mouth of Nizhnyaya Tunguska River and Ust-Port settlement. More reliable estimates of benthic biomass, 39.4 and 48.9 kg/ha, are given for the Yenisei delta and Yenisei Bay respectively (Greze 1957).

The Baikal sturgeon spends most of its life in the lake rather than in the rivers. Therefore, its growth is due primarily to the consumption of the benthos in the shallow near-delta regions and in the Baikal littoral zone (Yegorov 1961). In connection with the fact the most important for characteristic of sturgeon foraging conditions is the benthic biomass of the littoral zone up to 5 m depth. In the rocky bottom of the Baikal littoral zone, at 2-5 m, the mean annual benthic biomass is 67.42 g/m^2 (674.2 kg/ha). The seasonal oscillations in biomass are not great, ranging from 569.7 kg/ha in the winter to 828 kg/ha in the fall. In all seasons, by weight the benthos is dominated by molluscs (38.5-44.3%), gammarids (7.85-20%), and caddisflies (7.0-10.6%) (Kaplina 1974). In the southern part of Lake Baikal, in the zones influenced by runoff from the Baikal cellulose factory, there is a decrease in the biomass of oligochaetes (Kozhova et al. 1974).

In the Lena River, the principle foraging grounds of the most numerous lower Lena sturgeon population are the branches of the delta (Sokolov 1966b), as well as the side channels of the lower reaches of the river. The benthic biomass on various substrates of the Trofimovskaya branch in the Lena delta varies significantly: 2.2 kg/ha on silty-sandy substrates, 10.1 kg/ha on silty substrates, and 0.56 kg/ha on the whole for the branch. The main component of the benthos by mass is oligochaetes (82%) (Urban 1949). The mean benthic biomass in the main axis of the lower reaches of the Lena between Zhigansk and Sektyakh settlements, without taking into account the branches and side channels, is 0.36-1.96 kg/ha. The mean benthic biomass, calculated for whole lower reaches of Lena including its branches and delta, is 5.9 kg/ha (Vershinin 1964).

As can be seen from the data summarized in Figures 38 through 46 and in Table 31, the growth rate is generally correlated with (a) the thermal regimes and (b) the benthic biomass available as feed.

Table 31:
Net heat (sum of average daily temperatures in centigrade for the year) and benthic biomass in different sections of Siberian sturgeon habitat.

Water Body	Net heat (degree days)	Benthic biomass (kg/ha)	Source
Ob at Salekhard	1583	123.0-2165	Votintzev (1963)
Ob middle reaches - (Samarovskiy region)		9.79	Urban (1949)
Novosibirsk reservoir	2625.7	104.7-220.1	Orlova and Shirokov (1976)
Lake Baikal (0-50 m deep)	1400	569.7-828.0	Votintzev (1961)
Yenisei:			Greze (1957)
upper reaches	2222	-	
middle reaches	1789	10.6-11.5	
lower reaches		13.5	
delta		39.4	
Yenisei Bay		48.9	
Lena:			
lower reaches		0.36-1.96 without delta 5.9 including delta	Vershinin (1964)
Tit-Ary settlement	1079	-	Vershinin (1964)
Kyusyur settlement	1276	0.56	
delta	1348	-	Urban (1949)
Indigirka at Chokurdakh	1139	-	-
Kolyma at Cherskiy	1128	-	-
Kolyma at Zyryanka	1302	-	-

The fastest growing sturgeon population, that of the Ob River, because of two factors, (a) the highest amount of food availability and (b) the highest total annual heat flux. The rivers of Yakutia are coldest and have a less-developed benthos community and benthos abundance. Unfortunately, we do not have data on the benthic biomass of the Kolyma and Indigirka rivers. However, their thermal regime is similar to that of the lower Lena River. The lower Kolyma River is somewhat warmer than the lower Lena and Indigirka rivers. This explains most likely the somewhat higher growth rate of the Kolyma sturgeon with respect to the Lena and Indigirka sturgeon. The Yenisei River sturgeon has an intermediate growth rate compared to the Ob and Yakutian river populations. Correspondingly, the environment is also thermally intermediate and represents intermediate trophic characteristics. The most interesting population is the Baikal population. Figures 37 and 41 obviously show the growth rate of fish younger than 10-12 years, indicating that this is similar to that of the Yenisei and Kolyma populations. This is despite a significantly richer benthos community, although occurring in a somewhat colder environment than the Yenisei sturgeon. From about 11 years onward, growth sharply accelerates, and at ages 10 - 12 years, it quickly surpasses the growth rate of both, the Yenisei and Kolyma populations. At age 17 years, growth is faster than in the Ob population. The most acceptable explanation of this phenomenon is the previously described transition from a primarily benthic diet to an energetically more "efficient" piscivorous diet. An analogous acceleration in growth during such a switch in trophic feeding has previously been observed among Aldan River sturgeons (Sokolov et al. 1986). The transition of the Baikal sturgeon to piscivorous feeding is probably caused by the fact that, despite its extremely high benthic productivity, the littoral zone of Lake Baikal is very narrow and therefore the amount of available benthos for the sturgeon is insufficient, leading to sub-optimal food supply. This is supported by the fact that growth rate of Baikal sturgeon up to 10-12 years old (i.e. during the period when it feeds primarily on benthos), doesn't significantly differ from the growth rate of Yenisei and Kolyma sturgeon, which inhabit biotopes with much poorer benthic biomass (Yenisei) and lower temperatures (Kolyma).

3.3 Conclusions

The Siberian sturgeon exhibits an extreme plasticity in terms of its diet choice. The composition of it's diet varies significantly not only within the limits of species' range, but also with time (e.g. age or size). As typical benthophagous species, some populations of the Siberian sturgeon occasionally switch opportunistically and completely to a piscivorous diet.

Analysing the changes in diet composition along the rivers of various basins point to several general trends. In estuaries of rivers flowing into the Arctic Ocean, marine benthic fauna dominates in the diet. In deltas of the majority of these rivers - such as the Yenisei, Lena, Indigirka and Kolyma, sturgeons typically feed on amphipods and isopods. In the Ob Bay sturgeon feeds primarily on molluscs. In the rivers further upstream, the spectra of the diet widen and food composition changes significantly. Here the most important component are midge, caddisfly, mayfly, and stonefly larvae, as well as gammarids, molluscs and other invertebrate species.

In all studied populations, changes towards large sizes and amounts of food organisms occurred with increasing age and size of sturgeon. In certain populations (e.g. Ob River), we observed a diversification of diet with age. In the majority of the populations, (exception = Yenisei population) a partial transition to piscivory is observed at age 3-5 years. In some populations (e. g. Lake Baikal), older sturgeons feed primarily on fish.

A characteristic feature of the Siberian sturgeon is continuous feeding during winter over a large part of the range. Only in the Ob River sturgeon spawners stop feeding in winter. Probably the rich food resources in the Ob River basin allow the spawners to accumulate sufficient energy reserves for long-distance migration and spawning. The higher food available is indirectly indicated by the high growth rate compared to other populations.

With few exceptions, sturgeons in the Lena River do not feed prior to spawning. In the Indigirka and Kolyma rivers, feeding proceeds during the spawning season.

Seasonal changes in the diet content were observed in the majority of the populations of Siberian sturgeon studied. These data and differences between river basins and various river sections in diet composition support the existing opinion that sturgeons diet composition reflects the benthic species composition in the foraging grounds (Romanova 1948, Podlesniy 1955).

Growth in body length and weight is significantly different among populations and is directly related to the development of benthic food abundance in the respective water bodies as well as their thermal regimes. Growth rates for the Baikal sturgeon during benthophagy is similar to that of the Yenisei and Kolyma sturgeon, but increases radically after transition to piscivorous diets. Thus, the ecological plasticity of the species allows it to attain high growth rates even under conditions of insufficient food abundance for the majority of populations.

For the Siberian sturgeon, as in other fish species, there is a higher mortality of faster growing fish. As a result the maximal life span is found in slow growing individuals. In populations inhabiting severe conditions (Lena, Indigirka, and Kolyma), high longevity is with slow-growing females, which eventually attain also the largest sizes. This may be interpreted as adaptation to maintain a high reproductive capacity in areas of severe environmental conditions (e.g. low temperatures and food abundance).

4. Reproduction

4.1 Spawning habitats and time of reproduction

Siberian sturgeon spawn on gravelly-sand or gravel substrates at a depth of 4-8 m in currents of 2-4 km/hr (0.56-1.11 m/sec) (Podlesniy 1955, Yegorov 1961, Votinov 1963, Sokolov and Malyutin 1977, Ruban and Akimova 1991, 1993).

The spawning grounds in the Ob, are located primarily upstreams of the mouth of the Chulym River (about 2,540 km from the Ob river mouth). There are also spawning grounds in the lower Ob, including the region near the village of Nangy (Nadymskaya Ob branch) (Dryagin 1948b) and other places (Votinov 1963, Votinov et al. 1975). The spawning period varies but is generally from late May to early June in waters between 12°C and 18°C.

The Yenisei spawning grounds are located along a river stretch from the mouth of the Kureiyka to the settlement of Atamanovo (rkm 845-2,324) with the best sites occurring between the settlements Vorogovo and Sumarokovo (rkm 1,509-1,630 km). Spawning usually takes place between June and July at water temperatures of 16° to 21°C (Podlesniy 1955).

From Lake Baikal sturgeon enter the Selenga, Bagruzin, and Verkhnyaya (Upper) Angara rivers to spawn. Migrations begin in April before the ice has completely receded and continues to the middle of June. During this period water rises from 3-5°C to 14-15°C. The major migration takes place from the 5th -10th of May to the 5th -10th of June at water temperatures between 7 and 14°C. Spawning grounds are located in sections of the lower, middle and, more rarely, in the upper reaches of the rivers, for example in the Selenga River between rkm 22 to 155 where fish spawns from mid-May to mid-June (water temperature 9 to 15°C or higher) (Yegorov 1961).

The spawning grounds of the Lena sturgeon extend a great length from the upper reaches (Nuya River mouth and possibly higher) to the lower reaches of the river (Kirillov 1972), possibly even to the upper ends of the Lena delta (Pirozhnikov 1955). Spawning time differs between the northern and southern parts of the Lena basin. In lower reaches (at a well known site near the mouth of the Natara River (600 km from the Lena's mouth) sturgeon spawn from the middle of June to the middle of July (Akimova 1985a,b, Sokolov and Malyutin 1977). Our own observations identify a section adjacent to the mouth of the Olekma River, i.e. the middle reaches, as spawning area used between the end of May and early June (Koshelev et al. 1989).

For the lower Lena population, Sokolov and Malyutin (1977) report the approach of spawners to the spawning grounds at the mouth of the Natara (depth 6-8 m), at temperatures of 8-9°C followed by massive migrations once the temperatures reaches 12-14°C. As the freshet arrives, fewer sturgeon spawners are caught. The authors interpret this as a temporary slowing or even a pause in the spawning migration. However, according to our own observations, the snowmelt and rain-induced freshets in the Lena and other rivers in Yakutia are accompanied by a drop not only in sturgeon catches, but also in catches of other fish species. During a freshet, the rivers carry a large amount of suspended particles, plant detritus, timber, bush branches and trees, the water's turbidity is elevated. As the freshet recedes, however, fish catches, including sturgeon, increase to former intensity. Thus, it seems probable that a rise in water level leads to the dispersal of fish in larger water volume and their withdrawal from nearshore areas, where most of the fishing is done. Observations made over several years indicate that spawning migrations last about a month with three distinct arrival peaks of spawners at the mouth of the Natara: (1) middle of June, (2) end of June, and (3) the first third of July (Sokolov and Malyutin 1977).

Experiments have shown that egg development in the Lena sturgeon can occur in temperatures between 8 and 20°C while temperatures above 21°C causes embryo mortality. The optimal incubation temperature is between 11.4 and 14.9°C. Temperatures between 8° to 10°C and between 17 to 20°C are sublethal but result in noticeable embryo mortality (Nikolskaya and Sytina 1974). Other authors (Reznichenko et al. 1979) identify the optimal temperature range for of the Lena sturgeon egg incubation with 13 to 19°C. From fish hatcheries it is known, however, that incubation temperatures of 23-25°C produce quite viable offsprings from the lower Lena sturgeon (Sokolov and Malyutin 1977). Perhaps these differences have bio-technological reasons (e.g. use of different egg batches, inconsistency in the experimental setups, such as in the oxygen content of the water, current velocity and other factors).

To date, there have practically been no observations on spawning in the Indigirka.

Our investigations (Ruban and Akimova 1991) identified a spawning ground in the lower reaches of the river located on the right bank between rkm 306 and 315 km at the so-called Shamanovskiy steep bank, created by the jutting into the river of the Kondakovskoye plateau. The spawning ground is separated from the main channel of the river by the islands of Tatyanka and Shamanovskiy. Earlier Kirillov (1955) assumed a spawning ground near the settlement of the Druzhina (rkm 714). Appropriate spawning sites probably exist near the villages of Olenegorsk and Vorontsovo (rkm 327 and 363, respectively), where the right banks of the Indigirka are also rocky and high and the riverbed is gravelly.

In the lower Indigirka, sturgeon reproduces from July through the beginning of August in waters of 13°-16°C, as evidenced by catches of males undergoing spermiation (stage V) and stage IV females in late July.

In the Kolyma River, Siberian sturgeon reproduction has been virtually unstudied. However, our observations (Ruban and Akimova 1993) indicate that spawning grounds are located on a section of the river adjacent to the mouth of the Ozhogina River, a left tributary, about 900 km from the mouth of the Kolyma where many spermiating males and females with ovulated eggs and sexually ripe fish have been found. However, the many gravel bars in the Kolyma between the settlement of Zyryanka to the town of Srendnekolymsk (rkm 995 and 665 respectively) indicate that this section likely sustains other sturgeon spawning grounds.

Reproduction of sturgeon in the Kolyma takes place from the end of June to the end of July in water temperatures ranging from 16°C to 21°C, which is somewhat higher than in other rivers (Lena and Indigirka). We caught mature ripe females (stage IV) on July 24 (Ruban and Akimova 1993). It is possible that spawning continues in different years until early August. But histological analyses of gonads of female that we caught on August 3 with mature sexual products demonstrated pathological changes suggesting that the female had no possibility to spawn in time and oocytes will be resorbed. As a rule, males caught in July on the spawning grounds exhibit gonads at almost all stages of maturity, including V and VI, indicating that spawning is in process. In August, spent males and males with gonads in stages II, III, and IV were observed, the latter two indicate signs for getting ready to spawn the following year. Thus, males already have mature sexual products in the autumn of the pre-spawning year and overwinter in this condition (Ruban and Akimova 1993). This is a general rule for all male acipenserids.

4.2 Age at maturity, age and size of spawning stocks, and spawning periodicity

Tables 32 and 33 present data on age and size at which sturgeon have matured in different populations.

Maturation of Ob males occurs at a minimum age of 9 years (Votinov 1963). The females mature, according to some observations, at 9-12 years of age and according to others, at a minimum age of 16 years (Tab. 32, 33). In the upper Ob River, maturation seems to be reached somewhat earlier, at an age of 11 years. Spawning periodicity has been estimated to be every three years for males and once every five years for females (Petkevich

Table 32:
Size and age of Siberian sturgeon females at maturation and their reproductive periodicity for various bodies of water (a literature summary).

Water Body	Length (cm)	Body weight (kg)	Age (years)	Inter-spawning period	Source
Ob	103-110	9	9-12	4	Petkevich et al. (1950), Petkevich (1952), Dormidontov (1963)
Ob	-	-	16	4	Votinov (1963)
Ob	114-120	8.2	12	-	Dryagin (1947)
Irtysh	-	-	12	-	Menshikov (1936)
Irtysh	-	-	17-18	-	Bogan (1938)
Irtysh	-	15-17	16-17	-	Ereschenko (1970)
Yenisei	65-79	5-8	19-24	-	Podlesniy (1955)
Yenisei	85-90	4-6	-	-	Dryagin (1947, 1948b)
Lake Baikal	119-124	14	20-22	-	Yegorov (1961)
Lake Baikal	152-167	19-38	26-34	-	Afansyeva (1977)
Lena	80	1.4-2.0	-	-	Dryagin (1947)
Lena	70	1.5	16-20	-	Pirozhnikov (1955)
Lena	-	4.5	15-18	-	Karantonis et al. (1956)
Lena	70	1.5-2.0	11-13	3-5	Sokolov and Akimova (1976), Sokolov and Malyutin (1977), Akimova (1978)
Viluy	97	2.7	18	-	Kirillov (1972)
Indigirka	70-75	1.2-2.0	12-14	4-5	our data
Kolyma	79	2.15	16	4-5	our data
Konakovo warm-water hatchery	109	5.0-6.8	7-9	1.5-2	Akimova (1985a,b)
Rostov region ponds	103	7.2	10		Berdichevsky et al. (1983)

Table 33:
Size and age of Siberian sturgeon males at maturation and their reproductive periodicity for various bodies of water (a literature summary).

Water Body	Length (cm)	Body weight (kg)	Age (years)	Inter-spawning period	Source
Ob	103-110	9	9-10	3	Petkevich et al. (1950), Petkevich (1952), Dormidontov (1963)
Ob	-	-	8	3	Votinov (1963)
Ob	100	5.4	10	-	Dryagin (1947)
Irtysh	-	-	11-13	-	Bogan (1938)
Irtysh	-	-	8-9	-	Ereschenko (1970)
Yenisei	65-79	5-8	17-20	-	Podlesniy (1955)
Yenisei	75-80	2-3.2	-	-	Dryagin (1947, 1948b)
Lake Baikal	100	7	15	-	Yegorov (1961)
Lake Baikal	115-154	9.6-24	17-29	-	Afansyeva (1977)
Lena	70-75	1.2-1.5	-	-	Dryagin (1947)
Lena	60	1.0	15-18	-	Pirozhnikov (1955)
Lena	60	1.5-2.0	9-10	3	Sokolov and Akimova (1976), Sokolov and Malyutin (1977), Akimova (1978)
Indigirka	70-75	1.2-2.0	12-14	4-5	our data
Kolyma	79	2.15	16	4-5	our data
Konakovo warm-water hatchery	92-98	3.8-4.25	4+	1.5-2	Akimova (1985a,b)
Rostov region ponds	103	7.2	10		Berdichevsky et al. (1983)

et al. 1950, Petkevich 1952, Dormidontov 1963, Votinov 1963).

Maturation of the Yenisei River sturgeon has been reported to occur at 19-24 years of age for females and 17-20 years for males (Podlesniy 1955). The maximum age of the Yenisei spawners is probably rather high. In our catches, the age of females and males was not more than 49 and 51 years, respectively. Ages at maturation are similar for the Ob and Yenisei sturgeon populations (Tab. 32 and 33).

The males of the Baikal sturgeon attain maturity at about 15 years with a body length of 100 cm and a weight of 7 kg. Females mature later, at an age of 20 to 22 years, with a body length of 129 cm and a weight of 14 kg. Smaller mature females, only 7.2 kg, have also been recorded in the literature (Yegorov 1961).

Lena sturgeon females in the lower reaches of the river generally mature at age 11-13 years, with a body length around 70 cm and a weight from 1.5 to 2 kg. Males mature at age 9-10 (Sokolov and Malyutin 1977, Akimova 1978). In more southerly sections of the Lena basin, we observed much smaller spent spawners. Males from the Aldan River with ripe sexual products had a minimum size of 59 cm, a weight of 730 g and a minimum age of 9-10 years. Females with ripe sexual products had a minimum size 58 cm, weight of 890 g and a minimum age 10 years (Sokolov et al. 1984). In the middle reaches of the Lena, the minimum size of spent females was 56 cm, 735 g and 12 years (Koshelev et al. 1989). Histological analyses of the gonads of the Indigirka sturgeon showed that the majority of females attained sexual maturity at an age of 12-14, reaching then 70-75 cm in length, and 1.2-2.0 kg in weight (Ruban and Akimova 1991). These numbers are similar to those found in the lower Lena population. The range in age of the Indigirka sturgeon spawners is apparently more extended than in other populations. The maximum ages of reproducing females and males in our catches were 63 and 38 years, respectively.

Sexually mature Kolyma females at stage II of maturity with signs of having spawned the previous year (an accumulation of pigment granules in the connective tissue of the gonads) had the minimum size 79 cm and 2.15 kg at 17 years of age (Ruban and Akimova 1993). These data can be made more precise with more extensive sampling than were available for our use. However, it is likely that the size and age at maturity of the Kolyma sturgeon is comparable to that of the lower Lena River population (about 70 cm and 2 kg). The maximum age of spawning Kolyma sturgeon in our catches was 42 years for females and 30 years for males.

The cultivation of lower Lena sturgeon in warm water ponds and tanks (Tab. 33) has shown that females can reach maturity much faster once they had attained sizes typical of faster growing populations of the Ob and Lake Baikal.

Available literature data on the age composition of Siberian sturgeon spawners in the Ob-Irtysh basin cover the period from 1956 to 1970 (Votinov et al. 1975). In those years, the majority of the spawners were 20 to 39 years old, the individuals under 20 and over 40 years of age were less numerous.

As can be seen from Tables 34-37, the age distribution of males and females from the Ob-

Table 34:
Age distribution (in %) of mature Siberian sturgeon males from the Irtysh River (from Votinov et al. 1975).

Year	Age (years)							Mean age (years)	N
	9-14	15-19	20-24	25-29	30-34	35-39	40 and more		
1956	0.6	6.8	19.3	50.3	19.9	2.5	0.6	21.7	161
1968			2.3	38.6	40.9	15.9	2.3	41.4	44
1969		2.3	20.9	48.9	23.2	3.5	1.2	27.9	86
1970		1.2	8.5	31.7	41.5	14.7	2.4	30.9	82

Table 35:
Age distribution (in %) of mature Siberian sturgeon females from the Irtysh River (from Votinov et al. 1975).

Year	Age (years)							Mean age (years)	N
	9-14	15-19	20-24	25-29	30-34	35-39	40 and more		
1956		1.4	8.6	28.6	44.3	12.8	4.3	31.3	70
1968				4.8	33.3	46.0	15.9	36.2	63
1969			6.8	24.8	53.0	13.7	1.7	31.4	117
1970			1.6	12.3	41.9	33.6	10.6	34.5	122

Table 36:
Age classes (in %) of mature Siberian sturgeon males from the Ob River (from Votinov et al. 1975).

Year	Age (years)							Mean age (years)	N
	9-14	15-19	20-24	25-29	30-34	35-39	40+		
1956		1.4	15.7	57.2	24.3	1.4		27.9	70
1968			25.6	53.9	20.5			27.2	39
1969			20.0	32.5	32.5	15.0		29.1	40
1970			13.3	53.0	30.1	3.6		28.2	83

Table 37:
Age classes (in %) of mature Siberian sturgeon females from the Ob River (from Votinov et al. 1975).

Year	Age (years)						Mean age (years)	N
	9-14	15-19	20-24	25-29	30-34	35-39		
1956		1.4	5.4	28.4	47.3	17.5	31.2	74
1968			2.6	43.6	41.0	12.8	30.7	39
1969			3.4	17.2	48.3	31.1	32.3	58
1970			3.5	37.0	51.7	7.8	30.2	114

Irtysh system is rather broad. In the Irtysh River the spawning stock comprises more than 30 year classes of both sexes. Males are primarily 25 to 34 years, and females 25 to 39 years old.

In the Ob River, the age distribution of spawners is similar to that of the Irtysh but spawners older than 40 years are absent. In the Irtysh and the Ob the dominant year classes of females are between 25 and 39 years. The mean age of Ob-Irtysh males is 3.5-4 years younger than females. Votinov *et al.* (1975) determined the age structure of spawner and found this dependent on the number of separate generations and the intensity of exploitation.

Unfortunately, there are to date no comprehensive data on the age and size distribution of the Yenisei spawning population. With ages at first maturity for males and females of 17-20 and 19-24 years as well as maximum ages of 49 and 54 years (Podlesniy 1955), the maximum age range of the Yenisei spawners is 32-35 years. It is likely that this wide range of ages in the Yenisei is due to less intensive fishing than in the Ob River, allowing some older fish to survive.

The size distribution of sturgeon at spawning grounds in the Kolyma River has some unique features. As can be seen in Figure 48, the distribution is bimodal. Furthermore, the bulk of the individuals that make of the second peak (100 to 140 cm) consist of individuals with mature sexual products [III, IV and V stages of maturity]. In the lower reaches of the Kolyma, the distribution of sturgeon is unimodal, asymmetric and similar to the well-known size distribution of catches (Baranov 1918). Sexually immature individuals at the spawning grounds exhibit an analogous size distribution. It can be concluded that sturgeon in the Kolyma use various river sections, including spawning grounds, to forage. The bimodal size distribution around the spawning grounds is a consequence of the concentration of larger and older spawners superimposed on the typical foraging population. Similar bimodal age and correspondingly size distributions are observed in the Yenisei near the Vogorovo-Sumarokovo spawning grounds, where the spawning population is dominated by older age classes (Podlesniy 1955). A bimodal age distribution of the Yenisei sturgeon is observed between the Plakhino and Viski settlements, rkm 600-1,030, where the most common age classes are below 10 and above 25 years. According to Podlesniy (1955) this phenomenon serves as evidence of a so-called "anadromous" form of the Yenisei sturgeon, migrating from the spawning grounds to foraging grounds in the lower reaches of the river lasts up to an age of 20 years. However, on the contrary we believe that these data document the existence of a separate lower Yenisei population analogous to the one demonstrated for the Lena (see above). Arguments against Podlesliy's conclusions are supported by his own observations. Firstly, the late occurrence of the so-called "anadromous" form in the river up to an age of 20 years, which is almost the age when the fish reaches sexual maturity, and secondly, the existence of spawning grounds in the upper half of the section between the Plakhino and Viski settlements, above the mouth of the Kureyka (rkm 845 km from the mouth of the Yenisei). Thus, the data presented on a bimodal age distributions in sections of the Yenisei between the Plakhino and Viski settlements and

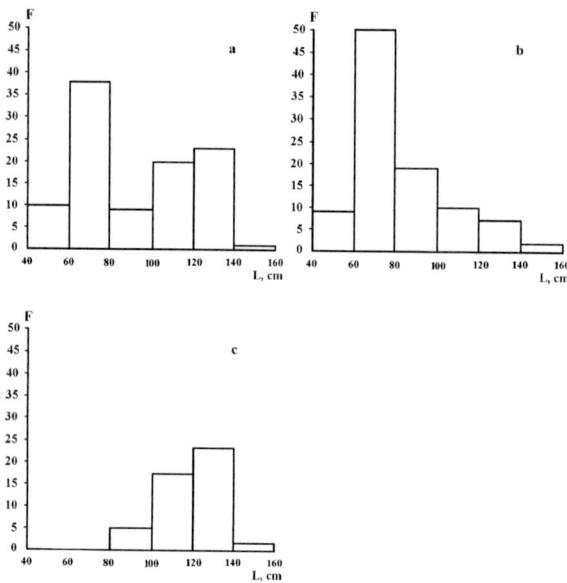

Fig. 48: Histograms of total body length distribution (TL) of Siberian sturgeon from the Kolyma River in catches on (a) the spawning ground 900 km from the mouth of the river, data obtained in 1988; (b) in foraging grounds 68-235 km from the mouth, data obtained in 1988-1989; and (c) mature reproducing sturgeon on the spawning ground 900 km from the mouth of the river, data obtained in 1988.

between Vorogovo and Sumarokovo indicates no more than the concentration of spawners around the spawning grounds.

Detailed observations on age and size composition of spawners in the Lena River exist only for the lower Lena population (Sokolov and Malyutin 1977). Here, as among the majority of sturgeon species, the females are larger than the males. The length of the females at the spawning grounds located near the mouth of the Natara River varies from 67 to 125.5 cm and their weight varies from 1,440 to 8,970 g. The bulk of the females (88.9%) compose a compact group from 75 to 100 cm in total length and 2 to 4 kg in weight. Spawning males range from 66.5 to 106.5 cm and 1,35 to 6,34 kg. As in females size and weight of the majority of spawning males (84.6%) fall within a narrow range from 70 to 95 cm TL and 1,500 to 3,500 g. The spawning part of the lower Lena population is represented by 20 year classes. The minimal age of mature males and females is about 9-10 and 11-12 years, respectively. Among the males, the more numerous (60%) are 13-16 years old. The bulk of the females (70%) are 14-18 years old. The maximum age recorded is 29 and 24 years for females and males respectively (Sokolov and Malyutin 1977). However, according to our own data, individual females can reach 50 years of age.

In the Indigirka River, the age distribution of spawners is wider than in the Lena River.

Spawning females range from 12-14 to 63 years and attain a weight of up to 22 kg.

In separate populations of the Siberian sturgeon, as had been shown above, a significant difference in life span for males and females has been observed. Populations having a high food availability are consequently characterized by higher growth rates (Ob, Yenisei, Baikal), and the maximal age of males and females is similar (Podlesniy 1955, Yegorov 1961, Votinov et al. 1975, our results). For sturgeons inhabiting the rivers of the Yakutia (Lena, Indigirka, Kolyma rivers), females, with a lower growth rate, following our data, have a greater life span. In populations with the lowest growth rates (Lena, Indigirka) life expectancy of females is roughly twice that of males (Tab. 30).

The ratio of males to females in spawning parts of all the examined populations of is close to 1:1, with a slight predominance of males (Podlesniy 1955, Votinov et al. 1975, Sokolov and Malyutin 1977). However, in separate generations of the Ob sturgeon, this ratio occasionally shifts and the predominance of males becomes significant to 2:1 (Votinov et al. 1975). According to our preliminary data, the sex ratios on the spawning grounds of the Indigirka and Kolyma are close to 1:1.

4.3 Fecundity

The absolute fecundity of Siberian sturgeon varies over a wide range (Tab. 38), for the species as a whole as well as for separate populations. The maximum value of 1,459,000 eggs, attained by Ob sturgeon, is directly related to that being the largest of the sturgeon populations. Records exist of Ob females caught weighting 192 kg and yielding 64 kg of caviar, equivalent to roughly 3-3.5 million eggs (Berg 1949, Votinov 1963). The relative fecundity of Ob sturgeon is comparable to that of other populations (Tab. 38).

The absolute fecundity of the Yenisei sturgeon is lower than that of the Ob sturgeon but is still relatively high. The absolute fecundity of the Baikal sturgeon is also rather high, and the relative fecundity varies within a range similar to that of the lower Lena sturgeon (Tab. 38). The diameter of the eggs ranges from 2.37 to 2.92 mm, their weight ranges from 10.9 to 16.1 mg.

The lowest absolute fecundity is seen among females in the lower Lena population and is probably related to the small size of the spawners. Their relative fecundity varies very widely. Minimum values of this character in the Ob, Yenisei and lower Lena populations are close, while it's maximum value in all studied populations are lower than in lower Lena population (Tab. 38).

The absolute fecundity of the Kolyma River sturgeon, similar to the Indigirka sturgeon but unlike the lower Lena sturgeon, is rather high and varies from 66 to 228 thousand eggs with female body weights ranging from 5.8 to 17.1 kg (Tab. 38). Earlier published results (Novikov 1966) on the fecundity of Kolyma sturgeon apparently in reality correspond to the Yenisei sturgeon (Podlesniy 1955). The high values of absolute fecundity of the Kolyma sturgeon, as for the Indigirka sturgeon, is related to the fact that in our catches females were mainly represented by larger individuals, in contrast to earlier analyzed individuals of the lower Lena population. Kolyma sturgeon females with mature sexual products are most often 8 to

Table 38:
Siberian sturgeon fecundity.

Water Body	Total body length (cm)	Body weight (kg)	Age (years)	Absolute fecundity (x1000 eggs)	Relative fecundity (1000 eggs/kg)	Sources
Ob		7-69.2		79.0-1459.0	9.0-24.0	Votinov (1963)
Irtysh	137-151	14-31	20-26	146.0-420.0	7.0-17.0	Yegorov (1961)
Yenisei		7-50		83.2-607.6		Podlesniy (1955)
	127-173		11-32	101.0-286.0	6.0-21.0	Yegorov (1961)
Lake Baikal	129-211	14-60	20-40	211.0-832.0	0.4-33.7	Yegorov (1961)
Lena	70-135	1.4-14.0		20.7-144.0	8.9-22.2	Sokolov (1965b)
		1.41-8.43	11-23	16.5-91.7	7.2-32.7	Sokolov and Ruban (1979)
Indigirka	115-141	10.05-21.0	33-58	105.56-245.34	15.0-17.0	Ruban and Akimova (1991)
Kolyma	105-137	5.8-17.1	18-42	65.6-227.84	9.34-24.49	Ruban and Akimova (1993)

12 kg, while at spawning grounds in the lower Lena females 2-4 kg in body weight dominate. The relative fecundity of Kolyma sturgeon females is similar to that of the lower Lena and varies from 9.3 to 25.5 thousand eggs/kg. The eggs of the majority of Kolyma sturgeon females are generally smaller than those of the lower Lena, reaching 11.5 to 18.2 mg, and 10.8 to 25 mg, respectively (Sokolov 1965b, Sokolov and Malyutin 1977, Akimova 1978, Sokolov and Ruban 1979, Ruban and Akimova 1993).

As can be seen from these results, the absolute and relative fecundity of Siberian sturgeon varies within very wide limits. Variability within a population is also quite high. For example, the absolute fecundity of the Yenisei sturgeon increases by a factor of almost 8 times with increased body size (Tab. 39).

The absolute and relative fecundity of lower Lena sturgeon females also vary greatly in connection with age and size (Tab. 40). In fact, the relative fecundity of the females drops with increased age and size, as is typical for other species of sturgeon (Sokolov and Malyutin 1977).

The data presented in Tables 32 and 40 show that all of the indicators that characterize the reproductive capability of females (absolute and relative fecundity, age, body weight) vary greatly for lower Lena sturgeon population. The weight of the eggs also exhibits great variability. Individual absolute fecundity is one of the most significant factors for characterizing the reproductive capability of a population. It is known that larger eggs produce larger offspring, which are considered to be more likely to survive (Gray 1928, Pertzeva 1939a,b, Meien 1940, Brown 1946, Morozov 1951, Hu Si-dzai 1957, Semenov 1963). In connection with this fact it seems to be actual to ascertain the connection between absolute fecundity and egg weight with female body weight, and also between absolute fecundity and egg weight.

We (Sokolov and Ruban 1979) have analyzed the relationships between these parameters using data collected by Sokolov during 1973-1978 on 264 lower Lena sturgeon females. Body length varies from 67 to 125 cm, weight varies from 1.41 to 8.43 kg, and age varies from 11 to 23 years. The results of the investigation (Tab. 41, 42, 43) show that absolute fecundity is strongly positively correlated to body weight (r=0.752, P<0.001), gutted body weight (r=0.684, P<0.001) and age of females (r=0.628, P<0.001).

Between body weight and egg weight (Tab. 44) there is a weaker but still significant positive correlation (r=0.36, P<0.01).

Thus, with increased size of females, their body weight, gutted weight and age, the absolute fecundity also increases. With increased body weight of the female the weight of the eggs also increases. It would seem, then, that greater absolute fecundity would be accompanied by greater egg weight. However, it turns out (Tab. 45) that egg weight and absolute fecundity are not correlated (r = −0.03, p > 0.05). This apparent independence is in fact the consequence of superposition of two relationships, conditioned on the qualitative biological diversity of the samples. In these investigations sturgeon females of 1,410 to 8,430 g body weight, aged 11 to 23 years were used, with an absolute fecundity of 16.5 to 91.7 thousand eggs. Clearly, this sample cannot be considered biologically uniform.

That is one reason to hypothesize that the relationship between egg weight and absolute fecundity would be more clearly shown in biologically more homogenous subsets, particularly of females of similar age and weight. Correlations between these two parameters were studied for a sample of 120 females for whom values of absolute fecundity, egg weight, body weight and age were known. For this sample, were calculated partial correlation between absolute fecundity and egg weight with excluded influence of body weight ($r_{12|3}$ = −0.39, p < 0.001), with excluded influence of age ($r_{12|4}$ = −0.54, p < 0.001) and with excluded influence of both weight and age ($r_{12|34}$ = −0.564, p < 0.001). As can be seen from these results, absolute fecundity and egg weight are inversely related, such that given an age group, or a size class, or both, greater absolute fecundity leads to lower egg weight.

Thus, greater size and age in Siberian sturgeon leads both to greater absolute fecundity and greater egg weight. But within age and size classes, that is, among individuals of the same age or size, greater absolute fecundity is related to smaller egg size. It is the combination of these two effects that explains the apparent independence between absolute fecundity and egg size for the pooled sample.

The relative fecundity of lower Lena population of sturgeon varies within a wide range, between 7.2 and 32.7 thousand eggs/kg, exhibiting a weak but significant inverse relationship with age (r = −0.17, p = 0.05) (Tab. 46) (Sokolov and Ruban 1979).

As can be seen from Table 46, the relative fecundity of sturgeons at ages between 12 and 14 years increases and stabilizes thereafter (14 to 18 years). In older age classes fecundity eventually drops to a minimum. It is known that the intensity of generative metabolism in the ontogenesis of fish increases until it stabilizes at a given age, in older age classes it drops off. A similar change is observed in the relative weight of the ovaries (dry

Table 39:
Yenisei sturgeon fecundity with size (from Podlesniy 1955).

Female weight (kg)	Absolute fecundity (1000 eggs)		
	Min	Max	Average
7-10	83.2	173.4	131.4
10-15	89.9	266.8	161.9
15-20	100.8	318.9	209.9
20-25	130.0	348.2	249.4
25-30	303.5	532.8	357.1
35-40	248.3	391.4	319.8
40-45	602.0	607.6	604.8

Table 40:
Absolute and relative fecundity among lower Lena females (from Sokolov and Malyutin, 1977).

Female indicator	Absolute fecundity (1000 eggs)		Relative fecundity (1000 eggs/kg)		n
Lenght (cm)	Range	Average and ± SD	Range	Average and ± SD	
60-70	16.5-28.0	22.3	16.3-30.1	23.2	2
70-80	17.2-46.1	29.5±1.2	9.9-37.5	21.3±1.0	37
80-90	21.2-61.9	37.7±1.0	10.5-33.7	20.0±0.6	75
90-100	26.9-81.0	50.6±1.7	10.8-31.9	18.9±0.7	48
100+	50.2-110.7	72.2±5.2	12.9-22.4	17.0±1.1	11
Weight (kg)					
1-2	16.5-35.1	26.6±1.4	15.3-30.3	22.0±1.5	14
2-3	17.2-54.8	33.7±0.9	9.9-37.5	20.3±0.7	74
3-4	25.5-64.2	44.1±1.2	10.5-28.9	18.7±0.6	55
4-5	40.6-81.0	58.3±2.7	12.6-31.9	20.2±1.1	20
5+	57.0-110.7	73.7±5.2	12.9-21.8	16.7±1.1	10
Age (years)					
11-12	16.5-46.1	27.6±2.5			11
13-14	17.7-54.8	34.4±1.6			34
15-16	21.2-61.2	37.5±1.4			46
17-18	22.8-74.1	43.1±2.0			41
19-20	35.4-81.0	50.9±2.6			17
21+	49.5-110.7	72.4±5.7			7

Table 41:
Correlation between absolute fecundity (1000 eggs) and weight for lower Lena females in 1973-1978 (Sokolov and Ruban 1979).

Absolute fecundity	Body weight (kg)								Total
	1-1.9	2-2.9	3-3.9	4-4.9	5-5.9	6-6.9	7-7.9	8-8.9	
15-24.9	3	7							10
25-34.9	4	23	7						34
35-44.9	1	18	19	2					40
45-54.9		5	13	7					25
55-64.9			5	4	3				12
65-74.9		1		2		1	1		5
75-84.9				1				1	2
85-94.9			1						1
Totals	8	54	45	16	3	1	1	1	129

Table 42:
Correlation between absolute fecundity (1000 eggs) and gutted body weight for lower Lena females in 1973-1978 (Sokolov and Ruban 1979).

Absolute fecundity	Gutted body weight (kg)								Total
	0.65-1.29	1.30-1.90	1.95-2.55	2.60-3.20	3.25-3.85	3.90-4.50	4.55-5.15	5.20-5.80	
15-24.9	1	8	1						10
25-34.9	2	21	9	2					34
35-44.9	1	10	24	5					40
45-54.9		5	9	10					24
55-64.9			3	7	1	2			13
65-74.9			1	2		1		1	5
75-84.9				1				1	2
Totals	4	44	47	27	1	3		2	128

Table 43:
Correlation between absolute fecundity (1000 eggs) and age for lower Lena River females in 1973-1978 (Sokolov and Ruban 1979).

Absolute fecundity	Age													Totals
	11	12	13	14	15	16	17	18	19	20	21	22	23	
16-21.9		3	1		1									5
22-27.9	1	1	4	2	5	2	3	1						19
28-33.9	2	2	1	4	6		4			1				20
34-39.9		1		4	7	5	4	1	1					23
40-45.9			2	5	2	3	3	5	3					23
46-51.9			1	1		3	4	1	2		1			13
52-57.9			1	1	1	1	1	3	3	1			1	13
58-63.9						1			2			1		4
64-69.9					1									1
70-75.9								2			2			4
Totals	3	7	10	17	23	15	19	13	11	2	3	1	1	125

Table 44:
Correlation between average egg weight (mg) and body weight (kg) for lower Lena River females in 1973-1978 (Sokolov and Ruban 1979).

Egg weight (mg)	Age (years)								Totals
	1-1.9	2-2.9	3-3.9	4-4.9	5-5.9	6-6.9	7-7.9	8-8.9	
10-12.9	8	7	3						18
13-15.9	13	41	17	6					77
16-18.9	7	51	35	10	4	1			108
19-21.9	1	11	16	11	3	1	1	3	47
22-24.9		5	2	2					9
25-27.9		1	2						3
28-30.9									
31-33.9			1		1				2
Totals	29	116	76	29	8	2	1	3	264

Table 45:
Correlation between absolute fecundity (x1000 eggs) and egg weight (mg) for lower Lena River females in 1973-1978 (Sokolov and Ruban 1979).

Egg weight (mg)	Absolute fecundity (x1000 eggs)								Totals
	1-1.9	2-2.9	3-3.9	4-4.9	5-5.9	6-6.9	7-7.9	8-8.9	
10.5-12.29		1		2					3
12.3-15.09		2	4						6
14.1-18.89		5	5	6	3	2			21
15.9-21.69	3	8	11	7	3	2			34
17.7-24.49		3	3	2	1				9
18.5-27.29	4	7	14	6	3		1		35
20.3-30.09	1	4	1	1	2	1	1	1	12
22.1-33.89	1	3	3		1				8
Totals	9	33	41	24	13	5	2	1	128

Table 46:
Correlation between relative fecundity (1000 eggs/kg) and age for lower Lena River females in 1973-1978 (Sokolov and Ruban 1979).

Absolute fecundity	Age (years)													Totals
	11	12	13	14	15	16	17	18	19	20	21	22	23	
7-8.9									1					1
9-10.9		1				1	1							3
11-12.9			1		1	1	3						1	7
13-14.9		1	1		2	1	2	2	1	1	1	1		13
15-16.9			2	1	4	8	1	5	1	3	1	1		27
17-18.9	1		2	2	7	2	4	2	3		1			24
19-20.9	1	1		4	2	6	1	3	2					20
21-22.9	1		2	1	1	1		1	2					9
23-24.9				1		2	1	2						6
25-26.9		2		2	1	1	2							8
27-28.9			2	1	1			1						5
29-30.9				2										2
31-33.9		1												1
Totals	3	7	10	17	23	15	19	13	12	2	3	1	1	126

ovary weight per gram of body weight) and relative fecundity, which is an analog to this indicator (generative metabolism) (Shatunovsky 1980).

An analysis of the Siberian sturgeon's gonadosomatic index (GSI), which reflects the level of its generative metabolism, has shown that over the past 30-40 years females of the Yenisei population have exhibited a decreasing GSI, whereas males have shown an increase of the index (Tab. 47).

A comparison of GSI in sturgeons from different water bodies shows significant inter-population differences (Tab. 48).

From Table 48, we see that at IV stage of maturity the greatest GSI values are found among females of the Lena population while the Ob and Yenisei fish have the smallest GSI. The GSI of the latter is similar to that of the Russian sturgeon from the Volga River. The Indigirka and Kolyma sturgeon females exhibit intermediate values. The observed inter-population differences in the magnitude of this index are most likely related

Part II - Ecological Characteristics of the Siberian Sturgeon

Table 47:
Changes in gonadosomatic index (GSI) in % in the development of sexual glands in the Siberian sturgeon and the sterlet of the Yenisei River (means and ranges). N = number of fish studied.

Stages of gonad maturation	Siberian sturgeon								Sterlet			
	Females				Males				Females		Males	
	our results		Podlesniy (1955)		our results		Podlesniy (1955)		our results			
	GSI	N	GSI	N	GSI	N	GSI	N	GSI	N	GSI	N
I	0.17 0.1-0.2	3	0.6		0.1 0.05-0.20	5	-		-		-	
II	0.9 0.2-3.6	4	1.4		0.7 0.1-2.8	31	0.4		2.5 1.5-3.6	5	0.9	1
III	-	-	9.9		5.1 2.0-10.1	3	2.9		-	-	4.6	1
IV	19.3 18.3-20.3	2	13.9-25.1		6.8 5.2-10.5	6	3.6-4.5		25.3 16.1-37.8	7	-	-

Table 48:
Gonadosomatic index (GSI) in % from stage IV of maturity among different acipenserids from various bodies of water. Means and ranges are tabulated.

Water Body	Females	N	Males	N	Source
Siberian sturgeon					
Yenisei	13.9-25.1		4.5		Podlesniy 1955
Ob	18.8-33.3		-		Dryagin 1949
Ob	20.0 11.3-30.0		-		Votinov 1963
Lake Baikal	23.6 16.9-31.5	31	-		Yegorov 1961
Lena	38.1 30.2-66.8	23	3.3 2.1-8.8	94	Ruban and Akimova 1993
Indigirka	26.59 24.2-28.79	2	-		Ruban and Akimova 1993
Kolyma	27.15 12.97-42.55	9	-		Ruban and Akimova 1993
Russian sturgeon					
Volga	25.2		-		Dyuzhikov and Serebryakova 1964
Volga	15.58-16.16		3.95-4.49		Pavlov and Pasporov 1971
Volga	23-28		-		Lukyanenko et al. 1974
Volga	20.6 13.2-23.9		-		Kazanski 1979
Volga	15.6-28.3		-		Veschev 1979
Volgograd reservoir	30 and more		-		Shilov 1964
Volgograd reservoir	-		3.9		Serebryakova 1964
Caspian Sea	-		9.3		Barannikova and Fadeyeva 1982
Siberian sterlet					
Yenisei	10.9-14.7		7.9-8.5		Khokhlova 1955
Ob	15.4-30.0		-		Dryagin 1949
Chulym	31.0 19.5-43.6		-		Usynin 1978
Volga sterlet					
Volgograd reservoir	18.2		3.5		Shilov 1964
Pyalovskoye reservoir (cages)	14.5		3.9		Mikheyev 1982

to differences in growth rate. Yenisei, Ob, and Chulym sterlet females in stage IV of maturation exhibit a similar GSI, which is greater than that exhibited by the Volga sterlet population.

4.4 Features of gameto- and gonadogenesis in Siberian sturgeon

Gameto- and gonadogenesis, from juveniles through maturation and subsequent reproductive cycles after the first spawning have been thoroughly studied by Votinov (1963) and Akimova (1978, 1985a,b), showing that these processes follows the same general pattern in Siberian sturgeon as in all other acipenserids. The specific difference in Siberian sturgeon relates to the timing within the reproductive cycle, determined both by age at maturity and the environmentally triggered periodicity. Some of these specifics are addressed below.

Gonads of Siberian sturgeon from the Ob River are weakly differentiated until the age of 4-7 years, making sex determination difficult. The initial reproductive cells and gonia begin to differentiate during in these age classes. The reproductive cells in three-year old females show oogonia only (Votinov 1963). Beginning with age 7, sex can be recognized visually. Ovaries have a well-expressed longitudinal groove along the lateral side. The oocytes enter the synaptic stage, their diameter attaining 11-13 µm. The subsequent phase of cytoplasmic growth is most extended, during which the size of oocytes increases from initially 20-25 µm to 460-500 µm. The structure of the nuclei, cytoplasm and envelopes changes during this period. The following phase the stage of vitellogenesis, takes also a long time among maturing females. While the yolk is being accumulated, the oocyte diameter increases from 550-600 µm to 2.5-3.0 mm. The age of female Ob sturgeon at end of the cytoplasmatic growth phase and the transition to trophoplasmatic growth is not indicated in the literature. However, taking into account the known minimal age of sexual maturity, about 16 years, and the duration of trophoplasmatic growth, which is no less than 2 years, we can consider that the period of cytoplasmatic growth takes about 6 years for maturing Ob sturgeon females (Votinov 1963).

The duration of the various stages of spermatogenesis of maturing Ob River sturgeon males are not indicated in the literature. Visually, the testes of four-year old males do not differ from the ovaries of females of the same age. The development of ampullas containing spermatogonia and seminal canals are observed in males over 4 years old. Spermatogenesis occurs asynchronously for the majority of males. At the beginning of the process separate canals contain reproductive cells in all stages of development - from dividing spermatogonia and spermatocytes of the first order to spermatids (Votinov 1963).

The unique feature of gametogenesis in Siberian sturgeon, which is related to the severity of the climate, the brevity of the arctic summer and the exhaustion of spawners during their spawning migration, is the increased duration of reproductive cycles relative to other species (Votinov 1963).

The duration of separate stages of development of reproductive cells upon maturation of the Siberian sturgeon females has been more precisely established for the lower Lena River population. Juvenile females can be up to 3-7 years old and

35.5-56.5 cm in length before they show signs of maturation. Females whose oocytes are in the synaptic phase are 3 to 8 years old, their length and weight varies between 47 and 67 cm and from 320 to 1,070 g. Beginning at age 5, some females enter the cytoplasmatic growth phase (Fig. 49). Females whose oocytes have entered the trophoplasmatic growth phase (Fig. 50) are no less than 7 years old. Upon attaining maturation, the whole vitellogenesis phase (stages III and IV) lasts no less than 4-5 years in females of the Lena sturgeon. The females of lower Lena sturgeon first spawn at 11-12 years (Akimova 1978, 1985a, b). However, the Lena population is also marked by a high variability in maturation rates and sizes of maturing females. As can be seen from the data above, some 56.5 cm females can be juveniles, but also at the same size can already have taken part in a spawning event (Koshelev et al. 1989). During repeated reproductive cycles, each phase of gonad development lasts at least a year (Akimova 1985a, b). On the whole, Lena sturgeon, as well as Kolyma and Indigirka sturgeon, are characterized by lower growth rates and earlier maturation at significantly smaller sizes than the sturgeon of the Ob, Yenisei and Baikal populations. The longer interspawning period among Siberian sturgeon, as compared to other acipenserids, is related to the shorter growing season and lower accessibility to food in northern water bodies than in southern ones, and respectively to longer periods of energetic resource accumulation are

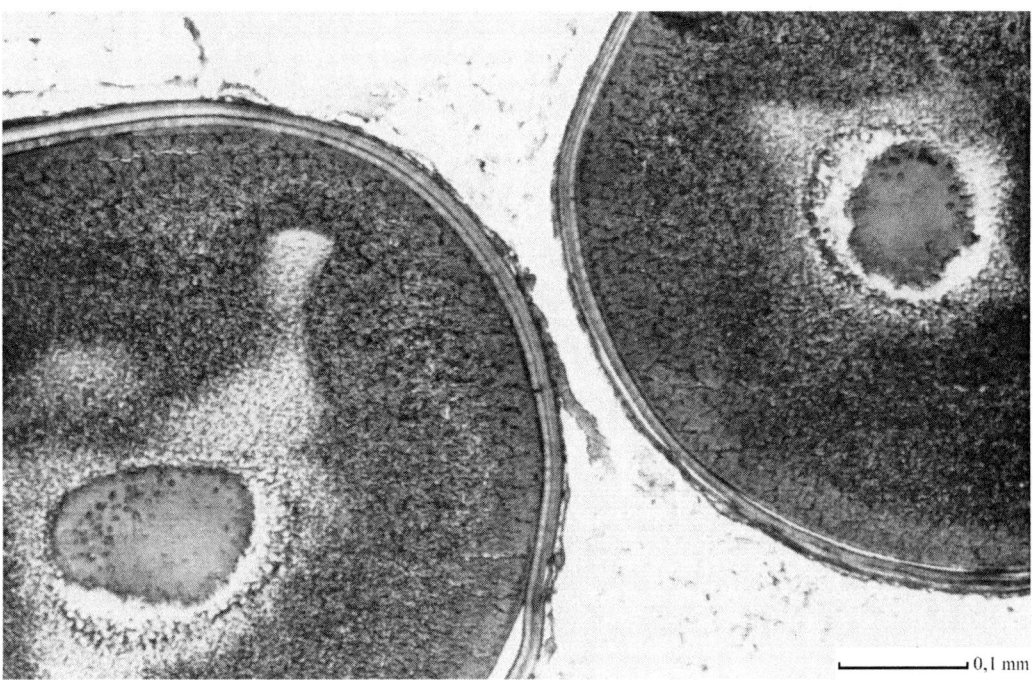

Figure 49: Microstructure of the gonads of Siberian sturgeon females from the Lena River in stages II-III of maturation; magnification: objective 8 × ocular 12.5.

Figure 50: Microstructure of the gonads of Siberian sturgeon females from the lower reaches of the Lena River in stage IV of maturation; body length 87 cm and weight 3,070 g; magnification: 8 × 7.

necessary for the development of reproductive cells (Votinov 1963, Sokolov and Malyutin 1977, Akimova 1978, 1985a, b). The cultivation of Siberian sturgeon in warm water environments is accompanied by a dramatic shortening of the time to maturation for males from 9-10 years to 3-4 years, and for females from 11-12 years to 7-8 years, that is, by factors of 1.5-3. Interspawning intervals under these conditions also shorten from 4-5 years for females to 1.5-2 years, and males can reproduce annually (Berdichevsky *et al.* 1983, Akimova 1985a,b). These results lend experimental support to the dependence of maturation rates and reproductive periodicity on trophic and thermal conditions as well as to the great plasticity of the species with respect to their reproductive rhythms.

4.5 Early ontogenesis

The features of embryonic development of the Siberian sturgeon have been insufficiently studied; though for the Lena population, the durations of separate phases of embryogenesis at different temperatures have been established (Tab. 49).

A comparison of the timing of various stages of embryonic development for the Siberian and Russian sturgeon at the same temperatures (Tab. 49 and 50) show that up to gastrulating, the development of the Siberian sturgeon is somewhat slower than that of the Russian sturgeon. Later, its development accelerates, and stage 18 occurs 5 hours earlier. Later stages of development for Siberian and Russian sturgeons (stage 26) occur simultaneously. The observed differences are insignificant and can be explained by incomplete correspondence of temperature regimes of the incubation processes of the study. It appears

Table 49:
Time of embryonic development for Lena sturgeon in hours from fertilization (from Malyutin 1965, 1980) at different water temperatures. The enumeration of the development stages is based on Dettlaff and Ginsburg (1954), Ginsburg and Dettlaff (1975) and Makeyeva 1992).

Stage #	Name of stage	Timing of embryonic development	
		11.3°C	14.3°C
Stage I: egg activation			
1	Fertilization	0	
2	Egg after turning inside membranes	2.5	2.0
3	Light crescente	4.5	3.0
Stage II: Cleavage			
4	1st cleavage division	6.0	4.5
5	2nd cleavage division	7.5	6.0
6	3rd cleavage division	9.5	8.0
7	4th cleavage division	11.5	10.0
8	5th cleavage division	13.5	12.0
9	7th cleavage division	16.0	14.5
10	Late cleavage division	18.0	16.5
11	Early blastula	23.0	21.0
12	Late blastula	34.5	25.0
Stage III: Gastulation			
13	Beginning of gastrulation	37.5	27.0
14	Early gastrula	41.5	29.0
15	Middle gastrula	52.5	33.5
16	Yolk plug of considerable size	58.5	35.5
17	Small yolk plug	69.5	45.5
18	Extended yolk plug	72.0	47.5
19	Early neurula	78.5	53.0
Stage IV: Embryonic development until beating of the heart			
20	Wide neural plate	81.0	55.0
21	Nearing of neural folds	83.5	57.0
22	Late neurula	85.5	58.0
23	Closing of the neuraxis	87.5	59.5
24	Appearance of eye growths	99.5	66.5
25	Nearing of the lateral plates	118.5	75.5
26	Fusion of the lateral plates	125.5	81.5
27	Short cardial channel	132.5	85.5
28	Straight, extended cardial channel	146.5	94.5
Stage V: Embryonic development from beating of the heart to hatching			
29	Development of cardinal vein	166.5	107.5
30	Tail approaches heart	174.5	
31	Tail reaches heart	189.5	
32	Tail touches head	201.5	
33	Tail goes behind head	215.5	

Table 50:
Time of embryonic development for Lena sturgeon in hours from fertilization at different water temperatures (from Dettlaff and Ginsburg 1954).

Stage	Name of stage	Timing of Embryonic development	
		11-12°C	14-15°C
13	Beginning of gastrulation		25.0
18	Extension of yolk plug	2.5	
26	Fusion of lateral plates	126	81
29	Development of cardial kink	149	

that the rates of embryonic development for Siberian and Russian sturgeons under the same temperatures are similar.

Studies on the postembryonic development of Siberian sturgeon show that the duration of the period from hatching to the transition to external nourishment (exogenous feeding) is about 15 days. On the whole, Siberian sturgeon development is rather similar to that of the Russian and stellate sturgeon, though there are some specific traits. Certain processes in the development of the digestive system occur after hatching in Siberian sturgeon, in contrast to the Russian and stellate sturgeon (Chusovitina 1963). For Russian sturgeon, stellate sturgeon, and beluga larvae, at hatching stage the internal endodermal part of the yolk sac wall is built by a symplast type, whereas for Siberian sturgeon larvae it is not a multi-nuclear symplast over its entire extent – rather, the entodermal part of the dorsal wall of the yolk sac is made up of single-layered epithelial cells. The cell plasma of the epithelial dorsal wall of the yolk sac in Siberian sturgeon larvae (after hatching) contains a large quantity of yolk pellets undergoing intracellular digestion, but the process of yolk capturing occurs neither immediately upon hatching nor later, when the epithelia becomes a symplast (Chusovitina 1963). In contrast to Russian sturgeon and stellate sturgeon larvae, the Siberian sturgeon larvae while hatching do not yet have a differentiated liver germ (Detlaf and Ginsburg 1954, Chusovitina 1963). Their liver germ, developed via the growth of internal entodermal parts in the front section of the ventral wall of the yolk sac, only becomes entirely differentiated in the larvae 3-4 days after hatching (Chusovitina 1963).

The rate of postembryonic development of Siberian sturgeon from the Lena population is greater than in Russian sturgeon, and the transition to external food under identical temperatures of the former occurs 2-4 days earlier (Yakovleva 1954, Malyutin 1980). A rather substantial feature of the postembryonic development in Siberian sturgeon is the nearly complete absence of feeding in the water thickness (between surface and bottom) by the larvae. The transition to the near bottom life style is made almost immediately after their yolk reserves are depleted (Malyutin 1980).

4.6 Conclusions

The Siberian sturgeon, like other acipenserids has a complex multi-aged population structure and relatively slow rates of recruitment. These species are well adapted to unstable conditions of reproduction. An isolated poor-reproductive year has little influence on the total population of a long-lived fish (Nikolskii 1974).

Acipenserids in general are marked by a lower lifespan of males (Nikolskii 1974). In this respect, the Siberian sturgeon is a rather plastic species. In populations living in more favourable conditions with high food accessibility and consequently higher growth rates (Yenisei, Ob), the differences in lifespan of males and females are not significant or non-existent. Other populations inhabiting more severe climates with lower food accessibility and consequently lower growth rates (Lena, Indigirka, Kolyma) are characterized by a much greater longevity of females, sometimes exceeding the lifespan of males by a factor of 2 (Tab. 30). This population structure, wherein the females are larger than the males, guarantees higher net rates of reproduction for the population and, respectively owing to the smaller size of the males, and greater under conditions of the same available overall food quantity for a population thereby maintaining its reproductive ability (Nikolskii 1974).

According the classification system of Lapin and Yurovitski (1959), the Siberian sturgeon's ratio for rate of growth in body length and age of maturity can be classified as type IV, characterized by slow growth and slow maturation. It is obvious that slow growth and extended maturation are rather relative terms. From Tables 32, 33 and Figure 36, it can be seen that different populations of the Siberian sturgeon differ significantly in terms of these parameters. There are populations with (a) relatively slow growth and accelerated maturation (Lena and Indigirka), (b) with fast growth and slower maturation (Baikal) and (c) fast growth with early maturation (Ob), which can be classified as type II, III or I, respectively. This diversity of growth and maturation rates is due to the wide variety of environmental conditions found over the extremely wide range of the Siberian sturgeon habitat.

Slow growth and accelerated maturation are characteristic for populations living along the edge of the distributional habitat range, i.e. living at the northern or southern boundaries of the species range. Albumen growth of fish in these populations is limited by the low accessibility of food and certain temperatures, and an early ontogenic shift of the ratio between albumin synthesis and fat accumulation in the direction to the last determines the accelerated maturation (Shatunovski 1980). Clearly, low level of food accessibility and unfavourable thermal regimes can be local and are not necessarily bound to the near border parts of the sturgeon's range, as is the case for the Lena and Indigirka populations which inhabit a colder climate with lower food availability.

The wide range in the relationship between growth and age at maturity among populations of Siberian sturgeon can be explained by the great plasticity of this species, i.e. its wide reaction norm. This is supported by the results of Lena river sturgeon rearing in warm water where elevated temperatures and abundant food supply leads to acceleration of growth by a factor of 7 to 9 and a lowering of age at maturity by a factor of 1.5 to 3 (Akimova 1988). Thus the population surpasses to the I type of these parameters ratio.

In Siberian sturgeon the increased size and age leads to increased egg weight and absolute fecundity. But within separated size and age groups (that is, among individuals of the same size and age) the egg weight drops with greater absolute fecundity. Relative fecundity varies widely for Siberian sturgeon, but generally peaks for young females between 12 and 18 years of age.

An analysis of inter-population differences according to relative fecundity and relative weight of ovary containing eggs of definite size (female GSI with gonads in Stage IV of maturity) has shown that the level of generative metabolism (for which these parameters are a characteristic index) is higher in slow-growing populations inhabiting areas with more harsh environmental conditions. This points to the adaptive plasticity of the species directed to maintain its numbers under varying levels of food availability.

Thus, as a species with a long lifecycle, the Siberian sturgeon is tremendously flexible in terms of size and age at maturity as determined by thermal and trophic conditions. In fast-growing populations with high food availability, the size of females attaining maturity is similar and near the maximum size for the species. The age of maturity under these conditions depends on thermal conditions and decrease with increasing of water temperature. In sturgeon populations inhabiting low-temperature and low food availability habitats (typical for fish in the northern edge of the habitat range), and also among slow-growing populations (Lena, Indigirka rivers), a physiological shift from somatic growth to generative metabolism is observed. This leads to earlier maturation of smaller individuals with an increase in the relative fecundity of the females. In these populations, the lifespan of males is less than that of females.

Embryonic development in Siberian sturgeon is possible over a wide range of temperatures. This is an important advantage considering the extended range of the species, inhabiting both arctic water bodies and warmer southern waters (Kazakhstan), as well as radically shifting regimes in the north where rapid fluctuations in temperatures in the summer are usual.

The significant features of early ontogeny for Siberian sturgeon is the accelerated development of the larvae in the postembryonic period in comparison to southern species of other sturgeons (Russian sturgeon), and the transition to near-bottom life style just after finishing the yolk sac stage. This can be interpreted as an adaptation of the Siberian sturgeon to the development in northern water bodies, characterized by a short growing season.

5. The state of reproductive systems in Siberian sturgeon populations and a proposed classification of reproductive disturbances potentially caused by anthropogenic impact

5.1 The influence of pollution on the development and functioning of the reproductive system

The anthropogenic impact on the natural environments, its various ecosystems, and individual species and their populations has been increasing during the past decades and its alarming rate is universally acknowledged. The overall data collected and results obtained on environmental trends and effects cannot be summarized within the scope of this monograph, even if it were just focused on aquatic ecosystems and on specific fish species. Therefore, the goal of this chapter is to briefly illustrate the linkages between disturbances in development and functioning of the reproductive system of fish and man-made pollution in the environment.

The main difficulty in assessing the degree of pollution influences on aquatic media and species is that there do not yet exist enough experimental data on the impact of specific pollutants on many of the species. However, it is known that threshold concentrations and type of influences, in particular heavy metals, on different fish species at different developmental stages differ significantly (Mur and Ramamurti 1987). The complexity of situation is potentiated by the fact that the cumulative action of toxins is largely unknown, with the net effects being unequal in different water bodies and vary with time, content and relative concentrations. Clearly, to study experimentally the effects of thousands of currently known toxic substances in all their possible combinations and relative concentration on hundreds of fish species is unrealistic. Thus, despite documented increases in pollution levels in most bodies of water, in practice it is almost never possible to confidently determine the relationship between the condition of a certain population and appearance of various deviations in development of organisms and concentrations of separate toxins. Consequently, data on pollutant levels in bodies of water (which is, incidentally, in itself rather dispersed and incomplete) has been of secondary importance in assessing the conditions of population and organisms. Instead, the condition of the organisms and their populations themselves is often provide the essential data in the assessment of environmental quality. This proposition is the foundation of one of the more modern approaches to the system of biological assessment of environmental quality known as the "Biotest" (Yablokov 1996). The essence of this approach consists in assessment of "ecosystem health" by integrating the assessments of the condition of the organisms that compose these systems. In this case for assessment of ecosystem health are in use the indices of organism's state, but not the global ecosystem or population parameters, the basic characteristic is homeostasis (Yablokov 1996). A disruption of developmental homeostasis is expressed in changes in the functional parameters of the organisms (Zakharov and Clarke 1993, Zakharov and Krysanov 1996). Among the many methods used in the assessment populations status and individual fish condition is the analysis of the development and functioning of the reproductive system, which is particularly sensitive to pollution (Akimova and Ruban 1992, Zakharov and Clarke 1993, Akimova 1996). The relevance of these methodological approaches is in the option to determine the quality of spawners, which is directly related to natural reproductive success at population level and, consequently, linked to prognosing recruitment (Nikolskii 1974).

The method of fish organism- and population state assessment, using the results of analyzes of reproductive system development and functioning, is based on the well known fact that significant changes in fish reproduction resulted by increase of influence of the whole complex of anthropogenic impacts (Koshelev 1949, 1981, 1988a and b, Lukshene 1978, Statova 1985, and others). The abnormalities of reproductive cell development among females of various species has been described by many authors: the partial degeneration of oocytes among cyprinids and percids (Kazanskiy 1949, Koshelev 1965 and 1981, Lukshene et al. 1979, Tatarko 1977, Beach 1959, Scott 1962, Chew 1973) and acipenserids (Burtzev 1962 and 1967, Faleeva 1979), amitosis among vimba (*Vimba vimba*) and acipenserids (Sakun 1965, Faleeva 1979, Akimova 1985a, b) and many others. We shall summarize some of the observations of anthropogenic impact on the reproductive system and reproductive success, largely ignoring the many hundreds of scientific papers on teleosts but mainly, paying particular attention to data related with the reproduction of sturgeon.

Among the various anthropogenic factors (physical, chemical, biological) impacting on fish, the role of toxic substances in particular has increased in recent years. Thus, in 1989, the pollution levels of about 40% of the monitored water bodies of were ten or more times higher than the maximum concentration limits (MCLs). Concentrations for many substances such as copper, zinc, nickel, and chromium compounds, highly oxidizing organic substances, oil products, phenols, pesticides and other toxic substances in tens of these water bodies exceeded the maximum allowable concentration level (MCL) by factors higher than one hundred (Pavlov et al. 1994).

Recently, many papers have been published presenting the results of environmental studies in natural water bodies and also those obtained in experiments. These report on the accumulation of toxic substances in various organs and tissues of fish, and their influence on the reproductive system of various fish species. Many researchers have discovered that toxicants in water not only influence the fish directly, but also have a particular impact on gameto- and embryogenesis leading either to the disturbance in reproduction itself or to the spawning of defective gametes, resulting in unviable offspring, thereby decreasing the potential for reproductive success of fish (Stroganov and Pozhitkov 1941, Stroganov 1970, 1971a, 1971b, Lesnikov 1970, Gusev 1971, Lukyanenko 1987a, 1987b, 1990, Mikhailova 1991, Pavlov et al. 1994, and others). It has been also shown that various herbicides, dispersants, petrochemicals and other toxic substances result in oocyte resorption in many fish (Bolkavdze and Gogotishvili 1988, Popova 1978, Koshelev 1988a, b, Selyukov and Stepanov 1988, and others). Heavy metal ions can have a negative influence on reproductive functions in carp *Cyprinus carpio* (Evtushenko et al. 1991). High concentrations of zinc cause profound abnormalities in gonad development and suppress fish reproduction (Alabaster and Lloyd 1984, Sehgal Rekha and Saxena 1986). Various forms of phosphorous in water can be highly toxic for fish (Krasnov 1970, Mazmanidi 1970). Disturbances caused by toxicants at the physiological-biochemical level were observed in sturgeons from the Caspian Sea basin, in domesticated and wild carp, silver carp *Hypophthalmichthys molitrix*, bighead *Aristichthys nobilis*, roach *Rutilus rutilus caspicus*, kutum *R. frisii kutum*, bream *Abramis brama*, chum *Oncorhynchus keta*, round goby *Neogobius melanostomus* and other fish (Vrochinskii and Zemkov 1978, Dokholyan et al. 1980, Anokhin and Muzyka 1988, Lukyanenko 1987, Geraskin 1989, Geraskin et al. 1989, Davletyarova et al. 1989, Shelukhin et al. 1989, Dzhavadova 1992, Natochin et al. 1995).

Many substances do accumulate in various organs of aquatic animals, particularly in the gonads (Zemkov and Zhuravleva 1987, Popova 1987, Gorkin 1990). Manganese, zinc, molybdenum, copper as well as pesticides can accumulate to extremely high concentrations in gonads, the liver and the adipose tissue of sturgeons, goldfish *Carassius auratus*, silver carp, and bighead during their pre-spawning period (Zubkova et al. 1989, Karimov 1989, Lukyanenko 1992). Concentrations of benzopyrene in molluscs and fish can be higher than in their environment, reaching more than hundreds or thousands times of the background concentration (Lembik 1976).

Toxic substances generally tend to accumulate in bottom sediments. Benthivores species, such as most acipenserids, are among the first who respond to benthic pollution (Kovaleva and Sergeeva 1987, Komarovski 1987, Komarovski et al. 1988, Alexandrov et al. 1988, Itra and Veldre 1988, Braginskii et al. 1989, Lukyaneko 1992, Sytnik 1992, and others).

Abnormalities in oocyte structure during the previtellogenesis manifest themselves in various ways. The initial changes can express themselves as deformation of oocytes during the cytoplasmatic growth and their karyoplasms vacuolization. The latter was observed in Russian sturgeon from the Volga River, and in sterlet from the Yenisei River (Shagaeva et al. 1993, Akimova et al. 1995a). This is probably related to the specific biochemical reactions of certain substances in the oocytes, since there are some data in the literature identifying dissolved substances (e.g. ions of sodium, phosphate and mercury) being able to penetrate easily into the fish bodies (Mur and Ramamurti 1987, Yarzhombek and Bekina 1987, Yarzhombek et al. 1991). During the pre-vitellogenesis in sturgeon females from the Caspian Sea basin changes in the cytoplasm structure have been observed upon exposure to toxicants. Morphological changes were highly correlated with concentrations in the gonads, muscles and liver of substances such as DDT, hexachlorcyclohexane, mercury and others (Romanov et al. 1990). Moreover, it has been shown that some substances, such as ammonium sulphate and malathion, cause a break-up of the nuclei and a delay in the development of gametes and gonad in the snakehead *Channa punctatus* and zebra fish *Brachidanio rerio* (Narayan and Sathyanesan 1986, Ansari Barde Alam, Kumar Kaushar 1987).

Over the past decades, direct nuclear division (amitosis) of oocytes during cytoplasmatic growth has been observed among sturgeons of the Caspian basin. Multinuclear oocytes and their amitoses have also been observed among Russian, great and stellate sturgeon during their foraging phase in the Caspian Sea. In 1981-1983, amitoses were recorded in 48.4% of the great sturgeon, 19.6% of the Russian sturgeon and 33% of the stellate sturgeon (Romanov et al. 1990). Studies of gametogenesis in the Siberian sturgeon from the Lena River in warm water aquaculture have also shown that changes of environmental conditions, followed by frequent appearance in females gonads of the amitotic gametes during cytoplasmatic growth. This phenomenon was not previously observed in the 1960's and 1970's in the Lena River (Akimova 1978, 1985a, b).

Earlier amitosis in various mammal tissues (connective tissues, bony and dense connective tissue, epithelial and liver tissue, male gametes, and others) has been observed and studied by a number of authors for many years (Zavarzin 1938, Zavarzin and Schelkunov 1954, Roskin and Struve 1948). Some authors consider this phenomenon normal (Karolinskaya 1952, Stroganova 1952). Others considered this a pathological feature caused by adverse environmental factors (Knorre 1959). In more recent studies, also carried out on mammals, amitosis was interpreted as a process of intracellular regulatory reaction, mostly associated with polyploidy (Brodskyi 1964, 1966; Brodskyi and Uryvaeva 1981). Almost all of these works mention nuclear division and only occasionally observe entire cell divisions. In fishes, amitotic division of nuclei and cells was described in a few papers on sexual cells of both males and females. Most authors consider this process to be a result of disturbances in cell development (Sakun 1954, 1964; Lemanova 1955, Faleeva 1979, Akimova 1985, Lavrova and Chmilevskyi 1987, Romanov et al. 1990). Chemical pollution is also considered to be one of the causes of this phenomenon in natural water bodies. The direct division of the gametes, following the opinion of some researchers, is an early defensive response by an organism, aiming to preserve it's species by multiplying the number of gametes in response to the deterioration of a habitat (Romanov et al. 1990).

Anomalies in oocyte structure related to pollution of water bodies manifest themselves not only in the period of cytoplasmatic growth, but also in later stages of development. Thus, the deformations of part of oocytes during cytoplasmatic growth in sturgeons described above have also been recorded during trophoplasmatic growth. The oocytes of indeterminate shape, so-called "deformed" oocytes (Sheveleva 1990), were found in the gonads in stage IV of maturity in Russian and stellate sturgeon females from the Caspian Sea. The deformation of the oocytes indicates a deterioration of turgor (ionic regulation) in the envelopes and a consequent loss of their strength. During the ingestion of nutrients into the oocytes, anomalies in their envelope structure were recorded. Local thinning of the jelly envelope and the presence of foreign inclusions within it as well as a exfoliation of the jelly and yolk envelopes, their uneven staining in microscopic sections, were observed in Russian sturgeon, stellate sturgeon and great sturgeon from the Volga-Caspian basin and in sterlet from the Yenisei River (Romanov et al. 1990, Shagaeva et al. 1993, Akimova et al. 1995a). Local thinning and exfoliation of the gamete envelopes also indicated the loss of their strength and uneven staining indicated that changes were occurring at a biochemical level as well.

Anomalies in the nuclear membrane structure in immature oocytes during trophoplasmatic growth were reported for female sturgeon from the majority of the studied populations. Thus 60% of the oocytes carried by 28.6% of the sterlet females from the Yenisei River showed abnormalities of this type (Akimova et al. 1995a). In great sturgeon females from the Ural River, 12% of the oocytes had abnormal nuclei. Females of Russian and stellate sturgeon foraging in the Caspian Sea had 8% and 4% eggs with this abnormality, respectively (Sheveleva 1990).

Beyond the abnormalities mentioned above, cavities containing a substance of unknown origin were observed in 1989-1991 in the mature oocytes of Russian and stellate sturgeon from the Volga River. These were located beneath the inner yolk envelope, between the jelly and yolk envelopes and among the yolk granules. In 1990-1991, the number and volume of these cavities were significantly higher compared to 1989 (Shagaeva et al. 1993). This abnormality has not been observed in oocytes of other species of sturgeon.

All of these abnormalities in gamete structure in different populations of sturgeon indicate a reduction of gamete quality and, consequently, of viability. Volga sturgeons have displayed an increasing tendency of these abnormalities and in changes of their type over the years. Thus, in 1988, the proportion of abnormal oocytes in different females of Russian sturgeon from the Volga River varied from 50 to 80%, in 1989 it varied from 90 to 100%, and in 1990 it was 100%. Stellate sturgeon in 1989 had 55-100% abnormal oocytes, 95-100% in 1990 and 100% in 1991 (Shagaeva et al. 1993).

For acipenserids as a whole synchronous oocyte development is typical, with negligible asynchronization only at the beginning of trophoplasmatic growth (Serebryakova 1964). However, under anthropogenic influences such as warm water aquaculture, the diameters of the trophoplasmic oocytes from a single female of the Siberian sturgeon differed by a factor of 2 to 3. In the largest oocytes, polarization of nuclei and yolk was already pronounced, whereas in the smaller oocytes it had not yet differentiated. It is likely that smaller oocytes are eventually resorbed, as is supported by the observation of remainders of pigment granules and basal membranes among the mature but not yet ovulated oocytes in the female gonads (Akimova 1985a, b). Volga sturgeon also exhibit asynchronous oocyte development during trophoplasmatic growth (Shagaeva et al. 1993). All of the examined populations of sturgeon species, both from the Caspian and Siberian basins, exhibited partial or occasionally massive resorption of the oocytes during cyto- and trophoplasmatic growth (Akimova and Ruban 1992, Shagaeva et al. 1993). As a result local accumulations of blood cells involved in the resorption process were observed. A large number of unspawned eggs often remain in the female gonads after the spawning period which also undergo resorption.

In recent years, various forms of hermaphroditism, previously unreported for sturgeon, have been observed (Romanov 1990, Romanov and Altufiev 1990, 1992, Romavov et al. 1990, Romanov and Sheveleva 1992). The proportion of fish exhibiting ovotestis reached 6.6% among stellate sturgeon and 2% among Russian sturgeon (Romanov et al. 1990). According to the authors, this condition can arise from disturbances in hormonal regulation due to the progressive accumulation of toxicants in organs and tissues. Hermaphroditism is rather rare among fish. It can occur in some species as a norm, for example, the *Sparidae* and *Serranidae* (Suvorov 1948, Alekseev 1969, Okada 1965a, b, and others). In whitefish *Coregonidae*, toothed carp *Cyprinodontidae*, ruffe *Percidae* and three-spined stickleback *Gasterosteidae* sex redifferentiation, one of the signs of potential hermaphroditism, can be triggered by changes in some environmental and ecological factors, such as an increase or decrease in water temperature, a change in diet, or a change in light conditions (Terian 1942, Chernyshova 1960, Anpilova 1965, Butskaya 1980, Berg and Hurk 1983). It has been shown that 8% of ruffe *Gymnocephalus cernuus* in Eastern parts of the Finnish Bay are potential hermaphrodites (Butskaya 1976), though this has not been observed in ruffe inhabiting waters closest to the bay. A discrepancy between the developmental state of gonads and gametes, on the one hand, and fish age, size and weight on the other (i.e. underdevelopment of generative part of the gonad) was another abnormality that manifested itself in the fact that individuals of adult age, size, and weight exhibited juvenile gonads.

Inclusions (neoplasms) of cross-banded muscle tissue in the ovaries have been observed in Russian sturgeon (*Acipenser gueldenstaedti*), stellate sturgeon (*Acipenser stellatus*), and great sturgeon (*Huso huso*). The proportions of fish with this type of pathology were 3.8% for the great sturgeon and 1.6-2.0% for the Russian and stellate surgeons. Tumours, cysts and other neoplasms on the surface of the gonads were evident in macroscopic investigations among 8% of Russian sturgeon females and 11% of the stellate sturgeon females (Romanov et al. 1990). The appearance of these neoplasms among Caspian sturgeons is presumably related to intoxication by carcinogenic compounds in the case of worsening of the physiological condition of the fish and the debilitation of the fish's immune defences (Romanov et al. 1990). Neoplasms were observed by the same authors in males of several acipenserids in the Caspian basin, including 7-8% of Russian sturgeon. Neoplasms have been seen on the gonads of whitefish (*Coregonidae*) in the Norilo-Pyasinskaya river system, which, for a long time, was strongly impacted by the intense industrial activities of the Norilsky metallurgical factory (Savvaitova et al. 1995). Furthermore, 9% of Russian sturgeon males and 10% of great sturgeon males showed swellings in the major connective tissue strands in the generative part of the gonad, i.e. the generative tissue of the gland was being replaced by dense connective tissue with a consequent depression in the development of reproductive cells (Romanov et al. 1990). A similar effect has been experimentally induced by exposing the reproductive glands of tilapia males to radioactive strontium (Voronina 1974). Silver carp *Hypophthalmichthys molitrix* in the cooling reservoirs at the Chernobyl nuclear power station in the post-catastrophe period exhibited seminal canals that were grown over by connective tissue with the creation of multiple cavities (Belova et al. 1993).

Reproductive cells (oocytes and gonia) have been observed in the liver tissue of Russian sturgeon, stellate sturgeon and great sturgeon in the Caspian Sea as well as during their migration in the Volga and Ural rivers (Romanov and Altufiev 1992). The authors suggest this may be due to the disorganization of embryonic cells by leaved the control inducers that determine the normal progression of embryogenesis for acipenserids during gastrulating and to a disruption of the conditions for the migration of the primary reproductive cells as a result of multi-factor toxic impact in early ontogenesis. Furthermore, they consider that: "... acipenserid 'adaptation' to pollution in the Caspian sea has reached it's limit and the defensive-adaptive mechanisms of the organism are no longer fit to provide reliable, normal functioning for individual cells and organs ..." (Romanov and Altufiev 1992, pp 153-154).

The presence of detergents in sewage runoff and consequent elevated concentrations of fluorine, bromine, iodine, and molybdenum ions as well as petrochemicals is blamed to induce a depression in spermatogenesis, shorten the longevity of the spermatozoids and lowers egg fertility in sturgeon, vimba *Vimba vimba* and loach *Misgurnus fossilis* (Kuzmina 1979, Stygar et al. 1979, 1981, Zimakov and Kuznetsova 1987).

A microscopic study of male gonads in Russian sturgeon form the Volga River in May 1990 revealed the presence of adipose tissue in the spermaducts (Shagaeva et al. 1993). This is not typical of properly developing male gonads. This appears to be related to a deviation in the lipid metabolism, as it is known that fish caught in heavily polluted waters exhibit high quantities of total lipids in their organs and tissues (Lukyanenko 1983, Savvaitova et al. 1995). Moreover, in 1990 was observed asynchronous spermatogenesis in several Russian sturgeon males form the Volga. While one section of the gonad contained fully developed spermatozoa, another sections contained gametocytes at different stages of spermatogenesis (Shagaeva et al. 1993). Replacement of some parts of the male gonads by liver tissue and vice versa was observed in sturgeon from the Caspian Sea basin during both their riverine and marine life-stages (Romanov and Altufyev 1992). First pathological changes in the gonads of sturgeon from the Caspian Sea basin were discovered as early as 1977 when abnormalities in gonad development were recorded in 2% of stellate sturgeon males from the lower Volga (Andronov 1983). These changes were manifested in different sturgeon species from this basin at different times, so that great sturgeon during their marine life did not exhibit any noticeable abnormalities at the early stages of gonad development (stage II of maturity) in 1982-1982 (Romanov et al. 1985).

Resistance to toxins during ontogenesis varies. Thus, mature sturgeon display less resistance than juveniles to organic toxins, but greater resistance to inorganic toxins (Lukyanenko 1967, Kokoza 1970).

Susceptibility to intoxication can be exacerbated by the warming of waters and by the consequent synergetic effect of exotoxicants on hydrobionts, and also by toxic, heat and radiation factors (Lukyanenko 1967, 1987a, Braginski and Kuzmenko 1988, Dedyu et al. 1988, Filenko 1988, Pomazovskaya et al. 1988, Abuelezz 1990a, b, Luksiene and Sandstrom 1994, and others). Moreover the toxicity of some substances, for example hexachlorane, can increase with lower temperatures (Zambriborshch and Bui-Lai 1976).

Radiation also causes significant abnormalities in gamete and gonad development in many fish, including bleak *Alburnus alburnus*, roach, carp, silver carp and tilapia *Oreochromis mossabicus* (Golovinskaya and Romashov 1955, Stroganov and Telitchenko 1958, Cherfas 1962, Voronina 1975, Voronina et al. 1974, Belova et al. 1993). Initial sexual cells (gonia and oocytes) in the early prophase of meiosis are most vulnerable (Zakharova 1983, Chmilevskyi and Zakharova 1987, Chmilevskyi and Ivoilov 1993).

Such disruptions in gameto- and gonado-genesis as degradation of the oocytes during cyto- and trophoplasmatic growth, the presence of a large number of residual eggs in female gonads after spawning, and the replacement of the generative part of gonads by connective tissue, all lead to a decrease of both absolute and relative fecundity in sturgeon females. It has been shown that from 1970 to 1985 the average individual absolute and relative fecundity of great sturgeon and stellate from the Volga River has decreased (Raspopov 1987, Veschev 1991). An experimental study on the impact of oil pollution at different concentrations on juvenile sturgeon revealed a number of effects that resulted in changes in individual behaviour, stunting of growth and a decrease in fecundity. Together, these effects can lead to the dramatic reduction of a species population and/or local extinction (Rustamova and Kasimov 1968). The replacement of the generative part of the gonads by connective tissue in males also resulted in a decrease in the number of spermatozoa released by males during spawning. The anomalies mentioned above as well as neoplasms in the gonads and the erosion of a part of spermaducts result either in an increase or decrease in the gonadosomatic index (GSI), but in all cases it follows by the detriment of the generative part of the gonads (Akimova et al. 1995a, b). After degeneration of sexual cells both during the cyto- and trophoplasmatic growth follows their subsequent resorption. The liquidation of these perished cells is natural process of organism protection. Resorption of the gametes can retard the development of the next generation of gametes and influence the rate of gametogenesis, as a rule, by extending certain periods. For example, the degeneration of oocytes during previtellogenesis and vitellogenesis of Pacific sardine *Sardinops sagax melanosticta* can cause a delay in maturation, a decrease in fecundity and the skipping of spawning (Sakun and Svirskii 1992). After mass degeneration of mature oocytes in acipenserids and resorption, the recovery of gonads can last a long time, prolonging the inter-spawning interval in some individuals. In light of these facts, it is clear that the most sensitive indicator of the level of influence of toxins on individual fish is the decrease of their fecundity (Suter et al. 1987), because the last is the final index on the value of which influence all numerous disturbances in development of the fish reproductive system both during their maturation and repeated reproductive cycles. However, the impact of these influences on population levels requires different detection methods.

During artificial production of sturgeons in various hatcheries high mortality among the larvae (80-100%), massive anomalies in their body structure, the absence or underdevelopment of the anterior section of the head, spinal curvature, dropsy of the pericardial cavity and yolk, malformations of the cardiac tube, etc. (Gorbacheva and Vorobyeva 1979, Igumnova et al. 1990, Shagaeva et al. 1993) were also observed. A number of these morphological abnormalities, such as deviations in morphology of fins, rostrum, caudal peduncle, barbels and spinal curvature have been reported in juvenile sturgeons from natural bodies of water (Kryazhev and Chebasov 1979, Shakhmatova 1983, and others). Their significance on the population level needs still to be quantified by adequate methods relevant for identifying population dynamics.

The results presented here by no means exhaust the amount of existing information on reproductive disturbances and serve only to illustrate the sensitivity of the reproductive systems among sturgeons to anthropogenic impacts, particularly to pollution. Furthermore, they show that deviations from normal functioning in the reproductive system can be a sufficiently reliable criterion for comparative assessment of population's reproductive state, the level of anthropogenic impact, and correspondingly indicates that studies to link these findings on individual level to the status at population level are urgently needed. These results are particularly important in situations where direct data on natural reproduction rates (number of spawners, efficiency of spawning, etc.) is unavailable, as is the case for Siberian sturgeon and may help to underline the need for such direct studies, without which no sound scientific proof on the true dimension (or contribution) of these effects to population decline can be made.

Thus, the study the reproductive systems in fish, which determine to a large extent the effectiveness of natural reproduction, is an perspective and urgent field of study in three interdependent aspects: a) a "traditional" examination of the relationship between reproductive potentials and durations of reproductive cycles on population; b) the identification of dependence of reproductive system state on separate environmental factors, directly related to problems of bioindication; and c) a determination of the degree of a given species prosperity in concrete environments, which requires a classification of disturbances in gameto- and gonadogenesis (Romanov et al. 1990, Akimova and Ruban 1992, 1996, Shatunovsky et al. 1996, Crespo 1990, and others). This last aspect of study is directly related to the analysis of mechanisms of reproductive system responses to various unfavourable impacts, including anthropogenic ones. Finally, all the three areas identified above need to be linked to quantifying studies on population dynamics.

5.2 The reproductive system of the Siberian sturgeon in Siberian rivers

Earlier studies on the development and functioning of the reproductive system in Siberian sturgeon have been undertaken only for the Ob and Lower Lena populations (Akimova 1978, 1981, 1985a, b, Votinov 1963). The main purpose of these studies, based on the data obtained in the 1960's and 70's, was to clarify the species-specific details of the normal development of the reproductive cells and glands. They do not contain information on deviations or pathologies in gameto- and gonadogenesis. Histological investigations of the reproductive systems in sturgeon from the Ob, Yenisei, Lena, Indigirka and Kolyma basins performed by us between 1982 and 1995 have shown that the overall morphological pattern of gametogenesis processes in Siberian sturgeon in these populations is similar to the general scheme which has been described for acipenserids from other water bodies (Akimova 1978, 1981, 1985a, b, Akimova et al. 1995a, b, Ruban 1996, Ruban and Akimova 1991, 1993). These studies have also shown significant abnormalities in the development of reproductive cells and organs over the past decades, as well as in previously studied populations of the Ob and the Lena rivers and also in populations that had formerly not been studied at all (Yenisei, Indigirka and Kolyma river systems).

Studies on the state of the reproductive system in sturgeons from natural bodies of waters (Lena, Indigirka, Kolyma, Yenisei, and Ob rivers) have shown that the number and spectrum of deviations from normal and the abnormalities in gameto- and gonadogenesis are not identical in different populations and vary with time

(Akimova and Ruban 1992, 1995, 1996, Akimova et al. 1995a, b, Koshelev et al. 1989, Ruban and Akimova 1991, 1993, Sokolov et al. 1986, Ruban 1994, 1996).

Long-term observations of the state of the reproductive system in the Siberian sturgeon from the Lena River have shown that between 1961 and 1977, there were only isolated cases of oocyte degeneration during cytoplasmatic growth, whereas by 1986, the ratio of females exhibited these abnormalities reached 59% (Fig. 51). Furthermore, have appeared other kinds of anomalies which were not detected earlier, e.g. the division of oocytes by ligature (hereafter referred to as "amitosis"), and degeneration of some oocytes during trophoplasmatic growth (Fig. 52), as well as high amounts of retained eggs after spawning in some females and rupture of the nuclear membrane during vitellogenesis (Tab. 51). Thus during long-term of observations was discovered the distinct tendency of increasing of number of abnormalities in gametogenesis and of afflicted individuals in the Lena population (Akimova and Ruban 1992).

The development of reproductive cells and glands in sturgeon of the Indigirka River had not been examined previously. Recent microscopic examinations of the female reproductive system revealed series of abnormal phenomena in the development of reproductive cells. Thus, deformation and partial or massive degeneration of the oocytes were observed in the beginning of cytoplasmatic growth in immature females (Figs. 53, 54). The oogonia do not degenerate. Oocyte degeneration is observed among fish of various age, size, and body weights. In 1984, 77% of the sampled females exhibited such disturbances; 90% in the sample from 1984; and 100% in 1987.

In histological preparations the process of degeneration occurs more than once in one female—there are a great number of fragments belonging to unstructured membranes of the follicular sheath, suggesting past resorption of a portion of the oocytes during cytoplasmic growth, in the egg-bearing layers of the ovaries. At the same time, degeneration of the next group of sexual cells may already begin (Fig. 55). The degenerative processes of gametes are noted also in fish of older age groups that have previously spawned. Thus of 16 large sexually mature females from 13 to 33 years with gonads at adipose maturity stage II, captured in 1987, normal development of the sex cells was recorded only in one (Fig. 56). In the others as well as in immature fish, either an extensive or partial degeneration of oocytes at cytoplasmic growth was recorded. In the connective and adipose tissues of gonads in some females, a great amount of dispersed granules of melanin pigments was seen, indicative of large-scale resorption of mature eggs from the previous year (Fig. 57).

Despite these pathological processes in the cytoplasmic growth period of the oocytes, the remaining normally developing oocytes continue their development and progress into the stage of vitellogenesis. However, even at this time, various anomalous phenomena are recorded—upon fixation and histological treatment, deformation of the greater share of the maturing oocytes occurs, which is an indication of the weakening of their envelopes. We also observe asynchronous development of oocytes during vitellogenesis, which is atypical for acipenserids and seem to indicate unfavourable environmental conditions (Akimova 1985 a, b). Furthermore, the remaining oocytes in the cytoplasmic growth stage that are found among the reproductive cells accumulating nutritive substances continue to degenerate. Some of the latter are also subjected to resorption. Thus, a single histological cross section can contain up to three resorbing oocytes in the trophoplasmatic growth stage. This is observed until the maturation of the sexual cells is completed (stage IV of gonad maturity).

After spawning, the females retain many unreleased eggs (Tab. 51) (Ruban and Akimova 1991, Akimova and Ruban 1992).

Thus, in the Indigirka population, as for the Lena population, the number of abnormalities in gametogenesis and of afflicted individuals increased, but over a shorter time interval and to a greater extent. Based on available data, the Indigirka females both immature and mature undergo massive degeneration of their reproductive cells during previtellogenesis. Though Indigirka sturgeon do spawn, the anomalies, distinguished at microscopic level, make it possible to suppose low quality of spawned eggs and additionally leads to the extended time of maturation of females and the longer inter-spawning intervals in connection with the continuous processes of degeneration of the reproductive cells, which takes place in the same female repeatedly.

The development and functioning of the reproductive system of Kolyma sturgeon had likewise not been studied previously. A microscopic analysis carried out by us revealed that the majority of females (81-83% of specimens under study) exhibited various abnormalities in their oocyte development (Tab. 52) (Ruban and Akimova 1993). In both immature and mature individuals in stages II of gonad maturation, deformations and degenerations of parts of the oocytes undergoing cytoplasmatic growth were observed. Amitosis, accumulations of blood elements and basal membranes of follicles, indicating the final phases of resorption, are present (Fig. 58). The main group of oocytes later begin to accumulate nutritive substances and these processes lead successfully to polarization of the yolk and to the migration of the nucleus to the animal pole of the oocytes – IV terminated stage of gonad maturity.

Despite these observations, there are also various abnormalities in oocytes structure during vitellogenesis, while some of the oocytes degenerate and are resorbed (Fig. 59). There can be as many as four such oocytes on a single histological slide. Some of the mature oocytes undergo thinning of the external (jelly) and internal (*zona radiata*) envelopes or local sticks or dents of the external membrane (Figs. 60, 61). All of this indicates a decrease of solidity of the envelopes, as a result of which 50 to 100% of the eggs have an irregular, angular form (Fig. 62). It can be presumed, then, that the majority of the Kolyma sturgeon in 1988-1989 spawned low-quality roe, the majority of which during further development could perhaps die. The similar thinning of ripe oocytes envelopes was observed among the Volga population of the Russian sturgeon in 1988-1990, which presumably can be a result of the poor quality of the Volga water (Shagaeva *et al*. 1989, 1991, 1993). There are reasons to believe that anomalies in the development of reproductive cells in the Kolyma River are likewise related to water quality. In the external (jelly) envelope of many mature oocytes of the Kolyma sturgeon inclusions of unknown origin were also observed (Fig. 63), a phenomenon that

Table 51:
Types of abnormalities in the gametogenesis of Siberian sturgeon in natural water bodies.

Type of abnormality	Lena 1986 n=33	Indigirka 1984-87 n=128	Kolyma 1988-89 n=119	Yenisei 1990-91 n=162	Ob 1995 n=5
Degeneration of oocytes during cytoplasmatic growth	+	+	+	+	-
Amitosis	+	-	+	+	-
Asynchronous development of oocytes during trophoplasmatic growth	-	+	-	-	-
Degeneration of oocytes during trophoplasmatic growth	+	+	+	+	+
Disruption of the nuclear membrane in oocytes during trophoplasmatic growth	+	-	+	-	-
Vacuolization of nuclear plasma	-	-	-	+	-
Asynchronous polarization of the oocytes	-	-	+	-	-
Egg deformation	-	+	+	+	+
Thinning of the jelly envelope	-	-	+	+	-
Cavities in the jelly envelope	-	-	+	+	-
Striation of the envelopes	-	-	-	+	+
Cavities within the yolk	-	-	-	-	+
Large quantities of residual eggs in the ovaries after spawning	+	+	-	+	-
Replacement of generative tissue with fatty tissue (fatty overgrowth of the gonads)	-	-	-	-	+
Neoplasms (tumors) on the gonads	-	-	-	-	+

Table 52:
Frequency (intensity) of abnormalities in the reproductive system of the Siberian sturgeon females from the Yenisei river (N = 46).

Type of abnormality	% females	% oocytes	
		Average	Range
Oocyte deformation	42.1	50.9	33.0-90.0
Oocyte degeneration	69.6	13.5	0.86-50.0
Occurrence of fragmentation in the basal membrane	84.8	6.2	0.20-18.60
Amitosis	47.8	0.7	0.10-2.20
Accumulation of the blood cells	45.7	—	—
Vacuolization of nuclear plasma in oocytes	23.9	2.1	0.40-5.0

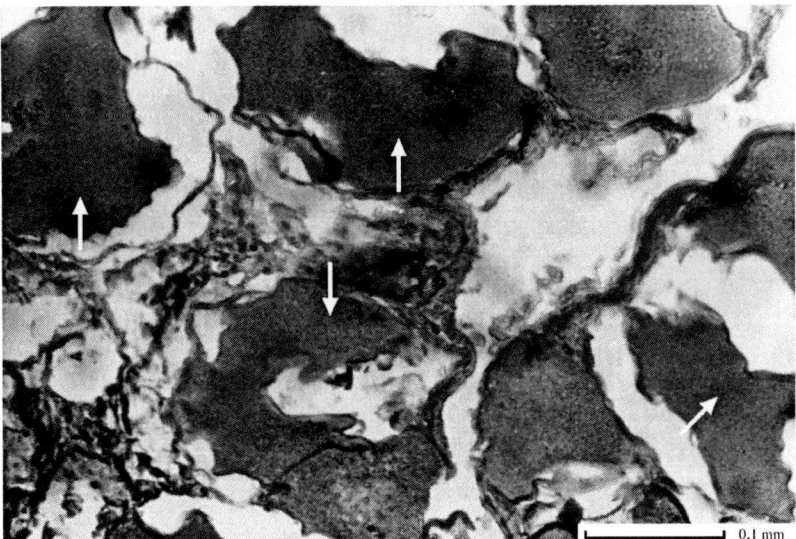

Figure 51: Microstructure of the gonads of Siberian sturgeon females from the Lena delta in stage II of maturation; total length = 64 cm; weight = 1.0 kg; age 8+ years; degeneration of a significant number of oocytes in cytoplasmatic growth (indicated by arrows) magnification: objective 9 × ocular 12.5.

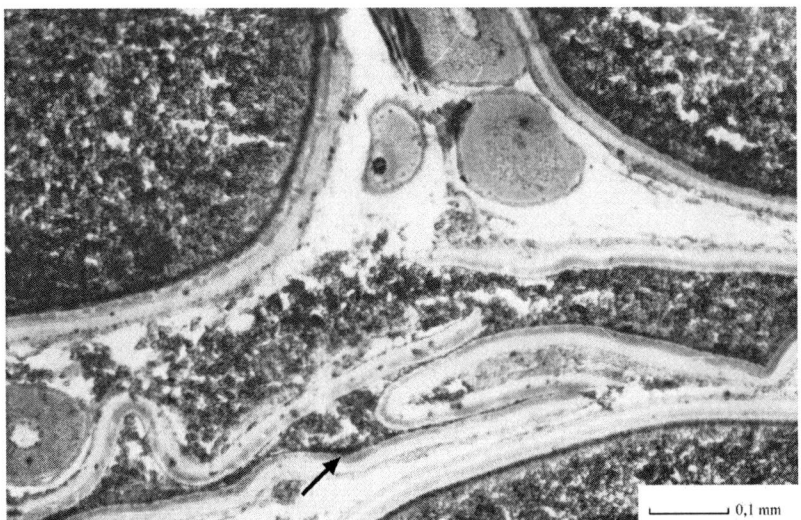

Figure 52: Microstructure of the gonads of Siberian sturgeon females from the Lena River delta in stage III of maturation; total length = 80 cm; weight = 2.08 kg; age 19 years; fragments from four oocytes can be seen filled with yolk, one of which is degenerated, magnification: 3.5×12.5.

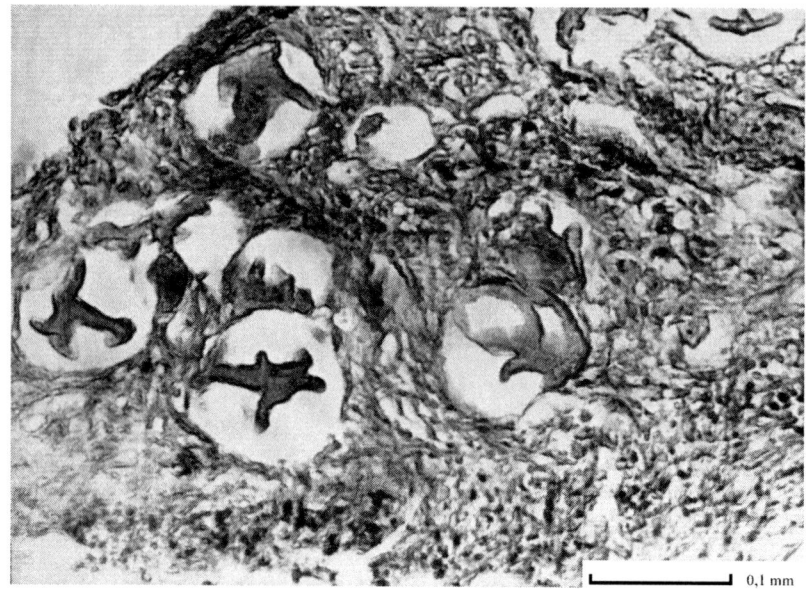

Figure 53: Degenerative processes in the reproductive cells in an immature female from the Indigirka River; stages I-II of gonad maturation; total length L = 38 cm; weight = 0.21 kg; age 4 years; degenerating reproductive cells entering cytoplasmatic growth (shown by arrows); magnification: 9×12.5.

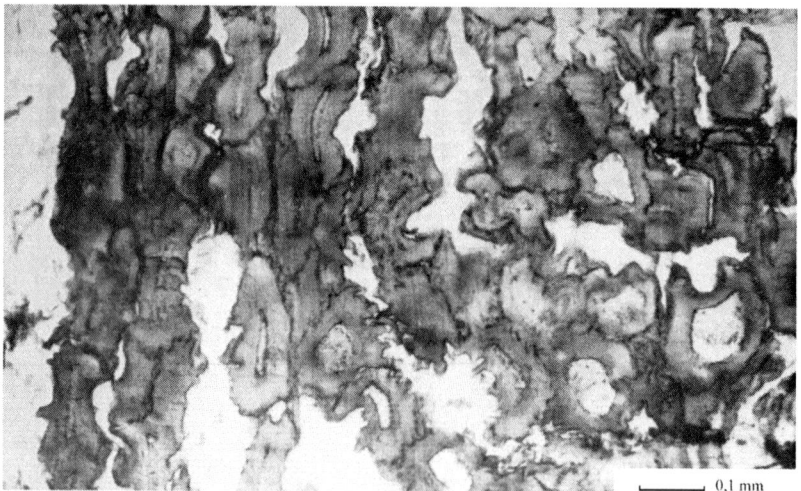

Figure 54: Degenerative processes in the reproductive cells in an immature female from the Indigirka River; stages II of gonad maturation, total length = 71 cm; weight = 1.55 kg; age 15 years; weight degeneration of oocytes during cytoplasmatic growth; magnification: 3.5×12.5.

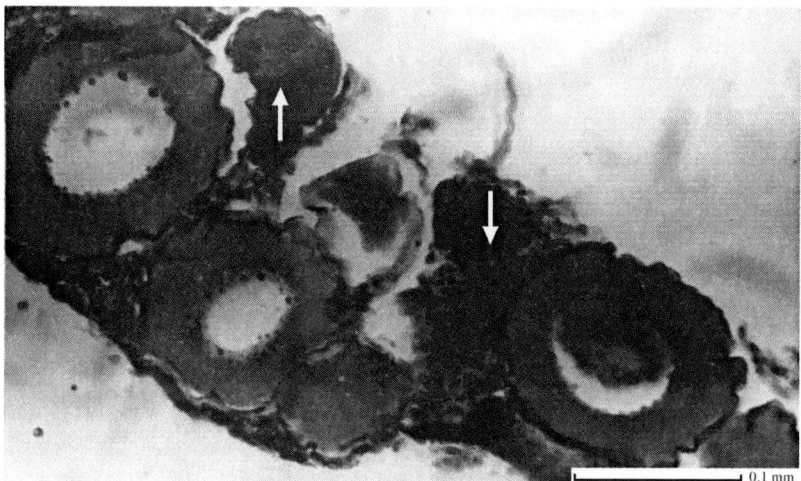

Figure 55: Microstructure of female gonads in Siberian sturgeon from the Indigirka River during previtellogenesis (stage II of gonad maturation); total length = 52 cm; weight = 0.495 kg; age 7 years; among normally developing oocytes, there are fragments of the yolk membrane of the follicular membranes of the oocytes as a result of past degenerations of reproductive cells; magnification: 9×12.5.

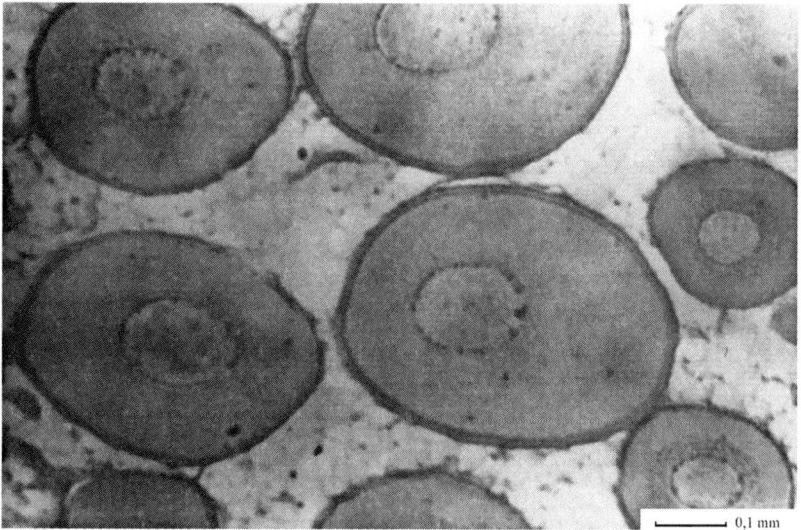

Figure 56: Microstructure of female gonads in Siberian sturgeon from the Indigirka River during previtellogenesis (maturation stage II of gonad maturation); total length = 112.4 cm; weight = 6.91 kg; age 31 years; normal development of oocytes in cytoplasmatic growth; magnification: 3.5×12.5.

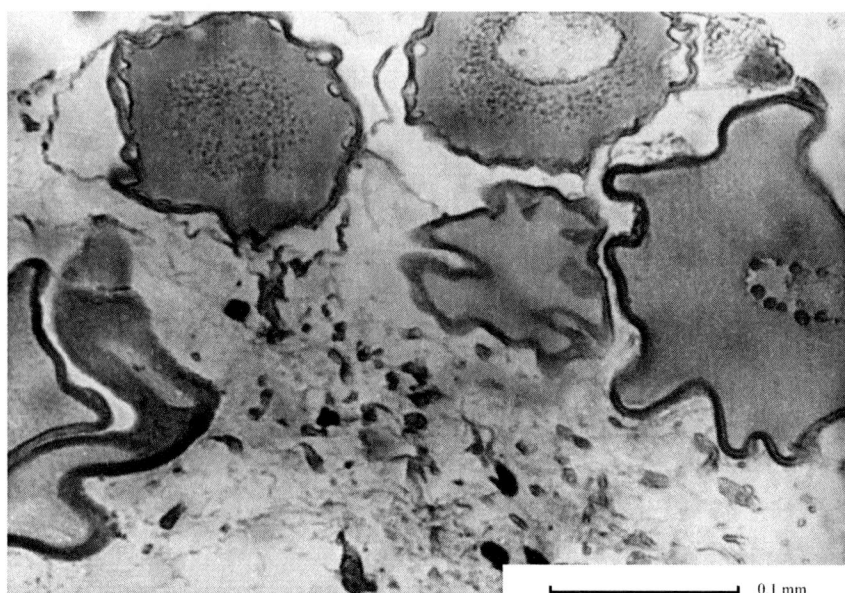

Figure 57: Microstructure of female gonads in Siberian sturgeon from the Indigirka River in stage II of maturation; total length = 99 cm; weight = 5.3 kg; age 13 years; granules of pigment in the fatty tissues of the gonads; magnification: 3.5×12.5.

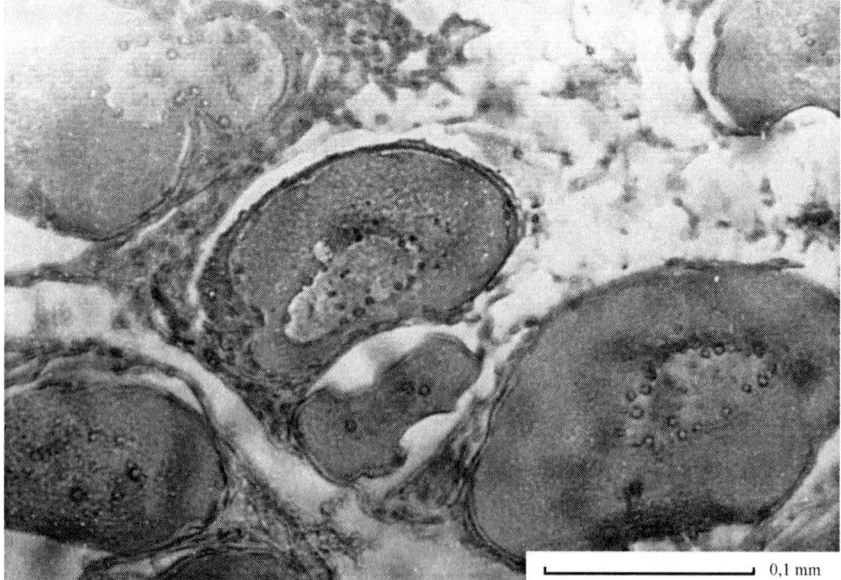

Figure 58: Microstructure of female gonads in Siberian sturgeon from the Kolyma River just at maturation (stage II of fatty gonad maturation); total length = 67 cm; weight = 0.95 kg; age 6 years; amitosis of reproductive cells and an accumulation of blood clotting; magnification: 9×12.5.

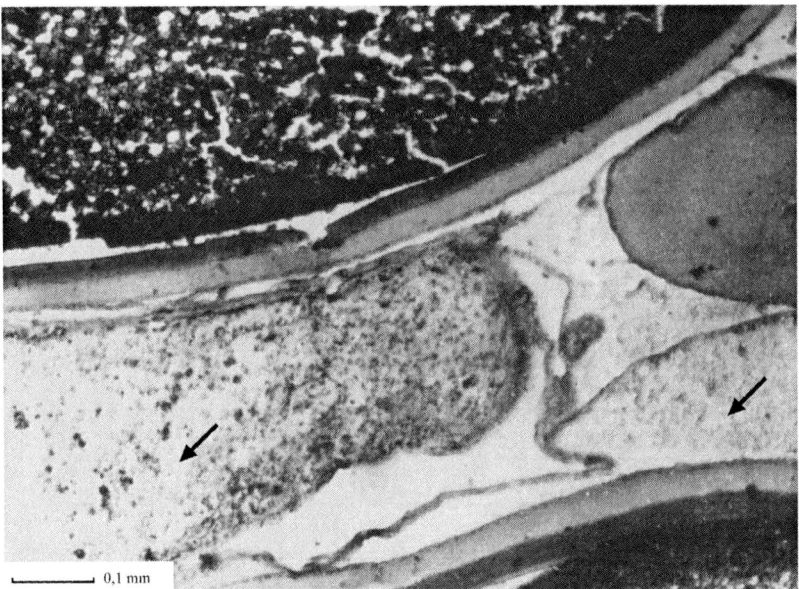

Figure 59: Culminating phases of vitellogenesis in the oocytes of a Siberian sturgeon from the Kolyma River (maturation stage IV); total length = 134.4 cm; weight = 14.88 kg; age 42 years; fragments from four oocytes in the period of accumulation of nutritional material, two of which are being resorbed (arrow); magnification: 3.5×12.5.

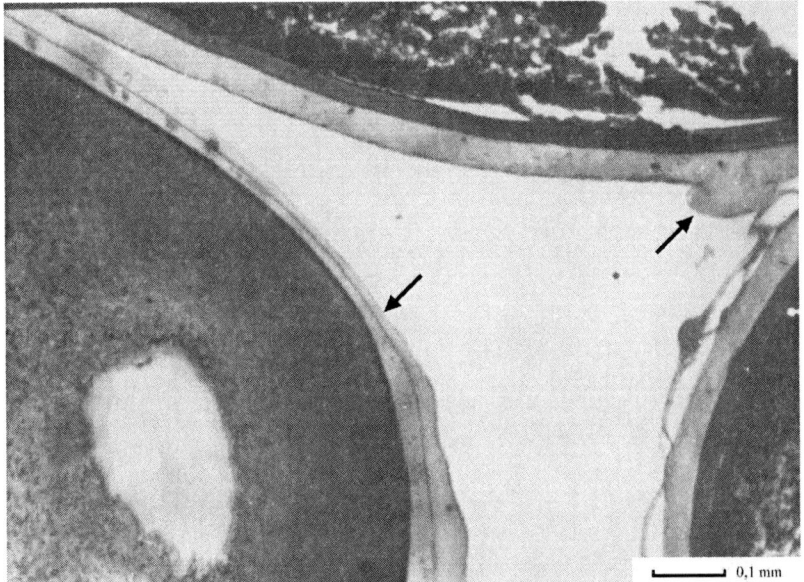

Fig. 60: Disturbances in structure of envelopes in ripe oocytes of the Siberian sturgeon female from the Kolyma river (IV stage of maturation, total body length 134,4 cm, body mass 14880 g, age 42 years, magnification: objective 3,5x ocular 12,5), thinning of the external (jelly) and internal (zona radiata) envelopes at the animal pole and local protrusion of the external envelope (indicated by arrows).

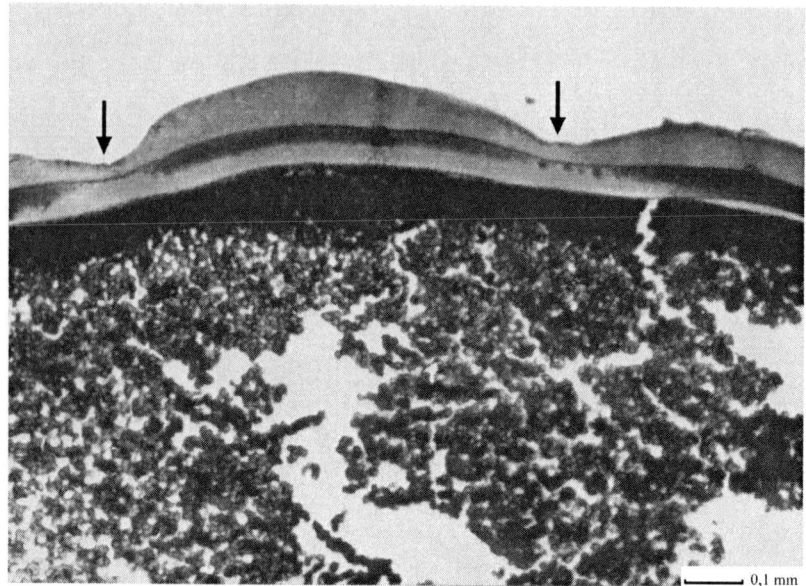

Fig. 61: Disturbances in structure of envelopes in ripe oocytes of the Siberian sturgeon female from the Kolyma river(IV stage of maturation, total body length 134,4 cm, body mass 14880 g, age 42 years, magnification: objective 3,5x ocular 12,5), local indentations of the external (jelly) envelope (indicated by arrows).

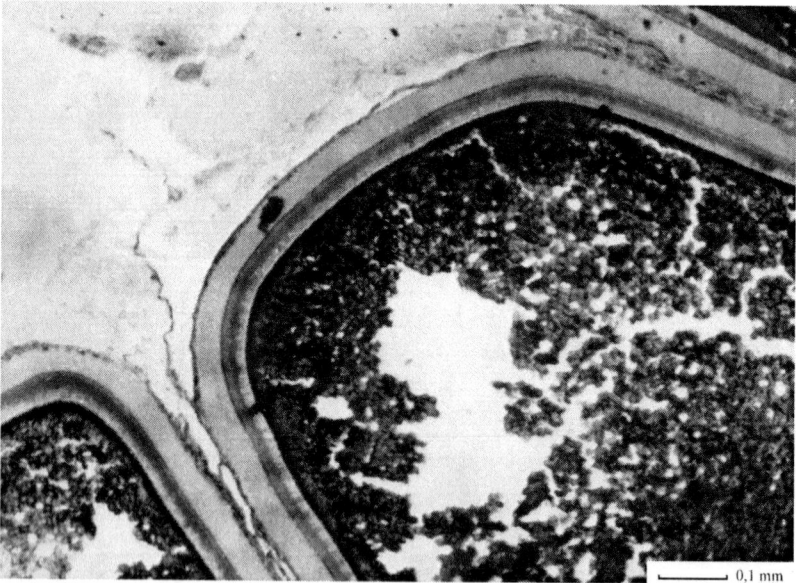

Fig. 62: Microstructure of female gonad in Siberian sturgeon from the Kolyma river in older age classes (IV-complete stage of maturation, total body length 155 cm, body mass 16100 g, age 30 years, magnification: objective 3,5x ocular 12,5), fragments of two deformed oocytes filled with yolk.

we have not observed in the samples from the Lena, Aldan and Indigirka rivers. Some females exhibit a destruction of the nuclei in oocytes that have not yet matured.

The overall picture of gametogenesis in the Kolyma sturgeon presents some unbalance in all processes - asynchronicity of vitellogenesis in oocytes (in one female, some oocytes had the nucleus in the center, in others polarization occurred), the absence of distinct borders between granules of large- and small-grained yolk (small-grained yolk extending "tongues" into the large-grained zone), and, as was previously described, degeneration of some oocytes during cytoplasmic growth and vitellogenesis. Degenerative processes in the gonads of the Kolyma sturgeon affect young oocytes to a lesser degree than in the Indigirka sturgeon; however, during vitellogenesis the picture is reversed.

Microscopic analysis of the reproductive organs of Kolyma males showed normal processes of spermatogenesis for the most part. However, some anomalies have been observed in the gonads of the males in stages II and III of maturation, including the disruption of the seminal canal walls and the appearance of cavities between the seminal canals (Fig. 64). Of the males in the sample that underwent histological analysis, 56% exhibited this type of abnormality. Analogous abnormalities were not observed among males of other Yakutian rivers (Lena, Aldan, and Indigirka).

The recent appearance of abnormalities in gametogenesis and the increase in the frequency of their appearances in up to 81-100% among females of Siberian sturgeon in the Indigirka and Kolyma rivers threatens certainly the natural reproductive success of the species. The most realistic explanation of the phenomena is the increase of water pollution resulting from intensified economic activity (increase of sewage and industrial runoff, development of river transportation, ore and other mining, etc.) since no other significant changes in the ecological conditions of these water bodies have occurred (climatic, hydrological, and those caused by hydro-construction).

The investigations on the reproductive system of Siberian sturgeon in the Yenisei delta in the winter of 1990 indicate that, there are several deviations from the normal development of reproductive cells which lower the individual female fecundity and the quality of the eggs, despite the externally apparent successful development and functioning of the reproductive system. During cytoplasmatic growth, degeneration of a portion of the oocytes and amitosis of reproductive cells occurs. During trophoplasmatic growth, local thinning and disruption of oocyte envelopes affects up to 50% of the egg cells in some females (Fig. 65), the appearance of inclusions of unknown origin under oocyte envelopes and among the yolk granules, and a mass resorption of ripe eggs among some individuals (Fig. 66) were observed. In the gonads of some spawned out females, large quantities of unreleased eggs remain. All these facts indicate the somewhat unfavourable state of reproductive system of the Yenisei sturgeon. The anomalies listed above are similar to those described from the Indigirka and Kolyma rivers and for acipenserids from the Volga River (Ruban and Akimova 1991, 1993, Shagaeva et al. 1993) but do not attain the same massive extent.

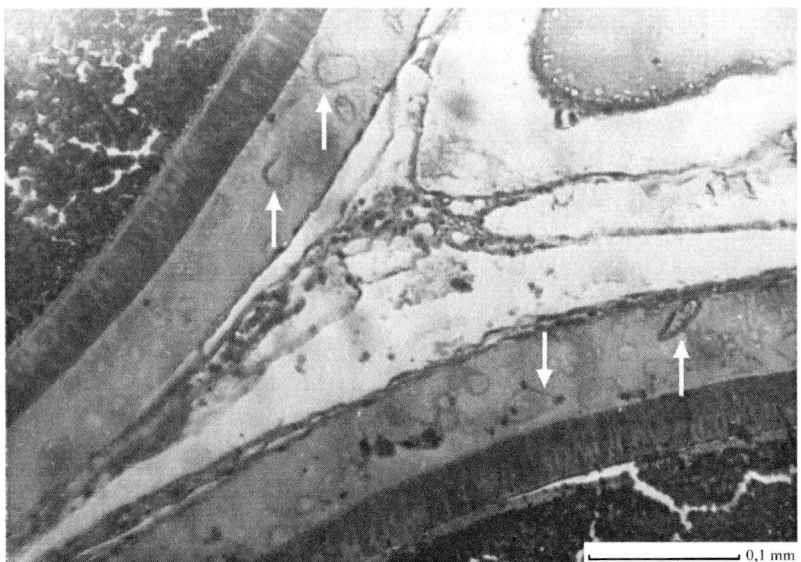

Fig. 63: Microstructure of female gonads in Siberian sturgeon from the Kolyma river (IV-complete stage of maturation, total body length 121 cm, body mass 10700 g, age 24 years, magnification: objective 9x ocular 12,5); fragments of two oocytes in the external (jelly) envelope in which foreign inclusions are visible (indicated by arrows).

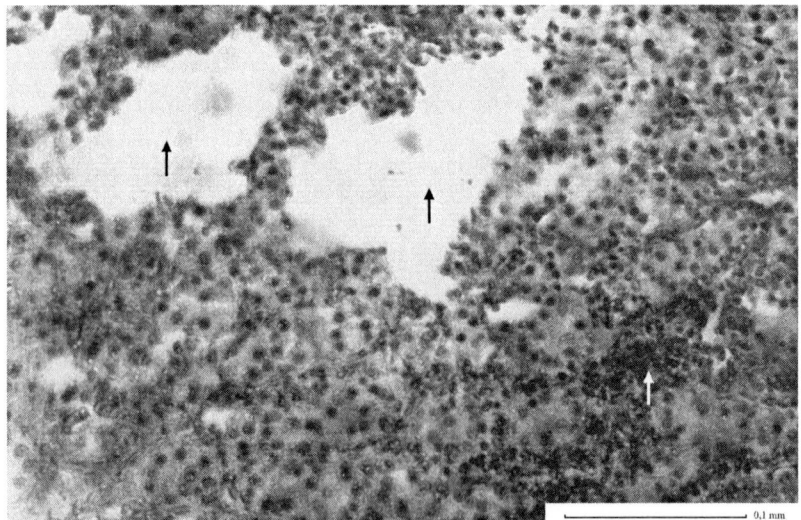

Fig. 64: Microstructure of gonads of a Siberian sturgeon male from the Kolyma river (II stage of maturation, total body length 102,5 cm, body mass 5250 g, age 23 years, magnification: objective 9x ocular 12,5); development of cavities between seminal canals (black arrow), blood clots (white arrow).

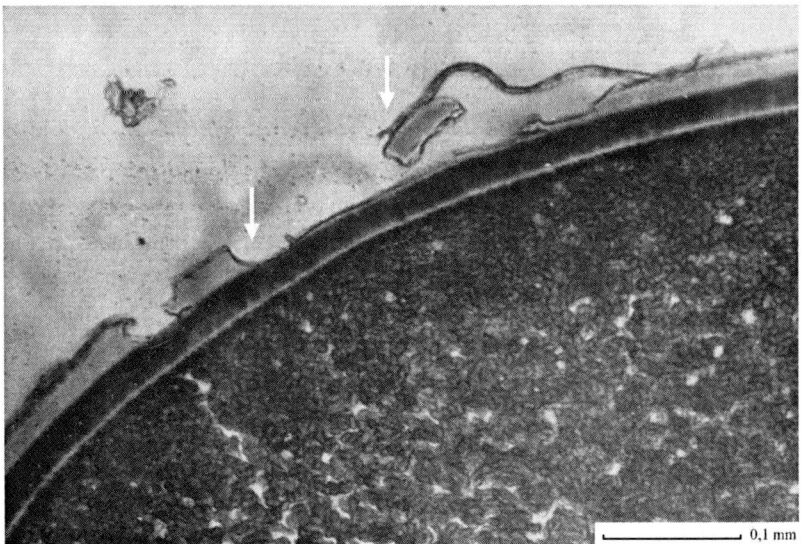

Fig. 65: Microstructure of the gonads of a Siberian sturgeon female from the Yenisei River in stage IV of maturation. Fragmentary destruction of the oocyte jelly envelope (indicated by arrows); magnification: objective 9x ocular 12,5.

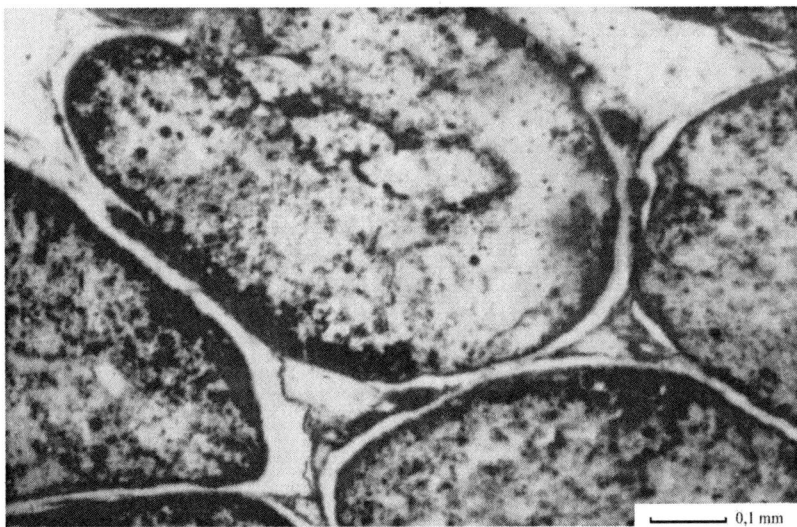

Fig. 66: Microstructure of the gonads of a Siberian sturgeon female from the Yenisei River in stage IV of maturation; mass resorption of ripe oocytes; magnification: objective 3,5x ocular 12,5.

Morphological investigations of the reproductive system in the lower Yenisei (20 km above the mouth of the Bakhta River) in August – September of 1990 and 1991 also displayed a variety of abnormalities in the reproductive system. Roughly half the females with previtellogenous oocytes had deformed oocytes (mean ratio is 50.9%). Roughly two thirds of the individuals undergo degeneration of oocytes in cytoplasmatic growth at a mean rate of 13.5%. These processes occur more than once in certain individuals as evidenced by basal membrane fragments that indicate past phases of oocyte resorption. These fragments are observed among the majority of females, their amount is not large and represent a mean of 6.2% of the oocytes in a cross-section (Fig. 67, Tab. 52). The gonads of roughly half the females exhibit accumulations of blood elements that participate in the resorption of oocytes (Fig. 68, Tab. 52).

During previtellogenesis, other abnormalities in the development of the reproductive cells were observed such as oocyte amitosis (47.8% of females) and vacuolization of the nuclear plasma. The quantity of Yenisei sturgeon females exhibiting these anomalies grew from 33.3 to 68.4% from 1990 to 1991. The mean number of cells undergoing amitotic division was 0.7% (from 0.1 – 2.2%) (Fig. 69, Tab. 52). These can have double and triple nuclei and a deformed shape (Akimova et al. 1995a). Subsequently, these cells probably also degenerate. Under normal development of reproductive cells and organs in fish, amitosis is not encountered. It has been earlier shown that amitosis in vimba (*Vimba vimba*) (Sakun 1964) and various acipenserids (Faleeva 1979, Akimova 1985a, b) is related to deviations in the reproductive processes.

Vacuolization of nuclear plasma in oocytes during previtellogenesis is likewise not encountered in normal fish development. First recorded in 1990, this phenomenon in Russian sturgeon from the Volga River during vitellogenesis (stage III of maturation) was associated with pollution in the river basin (Shagaeva et al. 1993). The percentage of Yenisei females whose oocytes were affected by these processes increased from 18.5% in 1990 to 31.6% in 1991.

There were only two females with gonads in stage IV of maturation in the sample from the lower reaches of the Yenisei River. Half of the oocytes in one female and all of the oocytes in the other were deformed. 90% of the ripe oocytes exhibited local thinning of the jelly envelope, exfoliation and disruption of the yolk envelopes and their vacuolization. In contrast to the incidences described by Faleeva (1965, 1979) in Russian sturgeon, these anomalies are not related to the mass atresia of ripe oocytes.

In males of the Yenisei sturgeon from the lower reaches of the river abnormalities in their reproductive system were also discovered. Localized destructions of seminal canals and appearance of cavities in the seminal tissue were observed in 68.8% of males in various age and size classes with gonads in all stages of maturity in 1990 and 43.8% in 1991. Large accumulations of blood elements, taking part in the resorption process, were observed near cavities in 35.4% of the individuals. Cavities among seminal canals have also been observed by us among males of the Kolyma population of Siberian sturgeon (Ruban and Akimova 1993) while these phenomena have not been observed in Lena, Aldan, and Indigirka

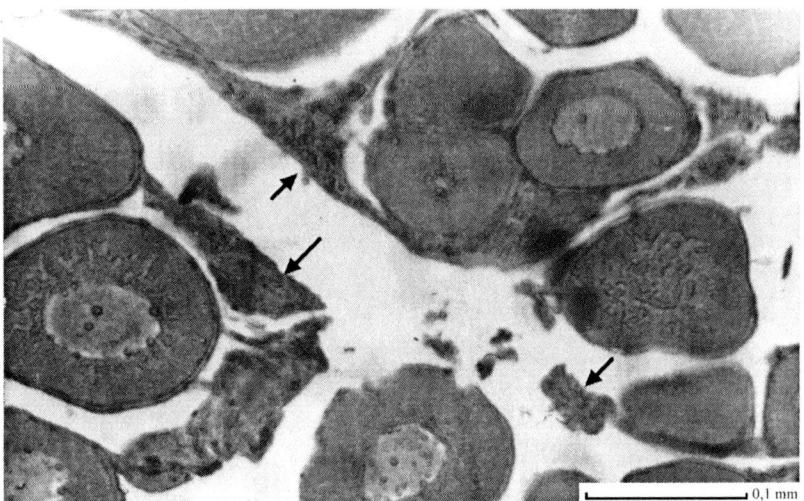

Fig. 67: Microstructure of the gonads of a Siberian sturgeon female from the Yenisei River during previtellogenesis (II-fatty stage of maturation, total body length 78.5 cm; body mass 1930 g; age 13+ years; magnification: objective 9x ocular 12,5); fragments of the basal membrane indicating the final stages of oocyte resorption during cytoplasmatic growth (indicated by arrows).

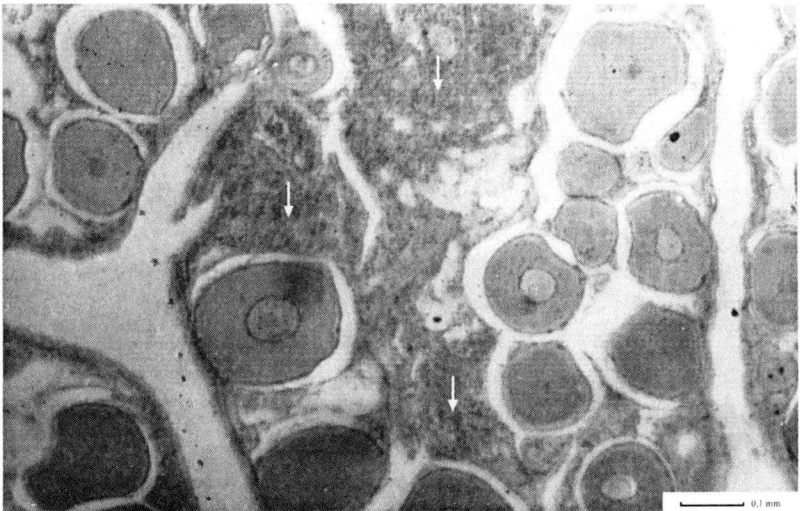

Fig. 68: Microstructure of the gonads of a Siberian sturgeon female from the Yenisei River during previtellogenesis (II-fatty stage of maturation, total body length 85 cm; body mass 3280 g; age 14+ years, magnification: objective 3,5x ocular 12,5); accumulations of blood elements taking part in resorption processes (indicated by arrows).

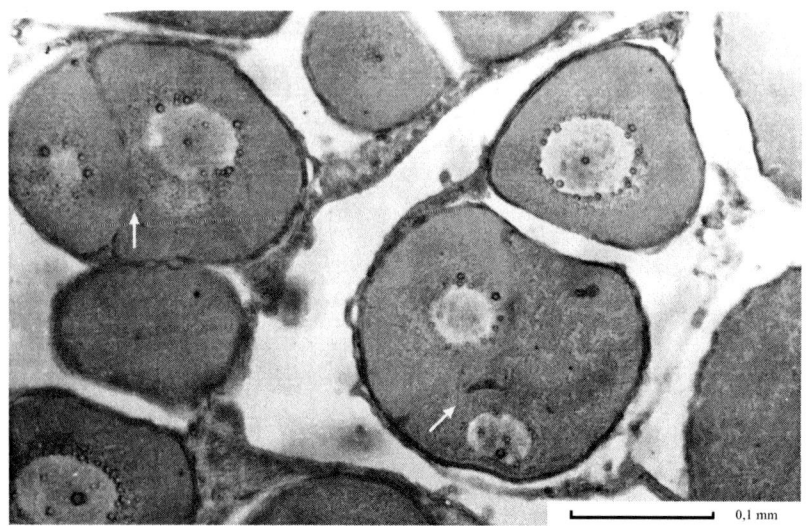

Fig. 69: Anomalies in the reproductive cells of a Siberian sturgeon female from the Yenisei River (II-fatty stage of maturation, total body length 104 cm; body mass 4300 g; age 15 years, magnification: objective 9x ocular 12,5); amitosis of reproductive cells during cytoplasmatic growth (indicated by arrows).

populations (Sokolov et al. 1986, Koshelev et al. 1989, Ruban and Akimova 1991). Furthermore, 4.2% of the males exhibit lobularity of the testes reminiscent of the formation of egg-bearing laminations in females, and 22.9% of the males exhibit swellings of the major connective tissue strands in the generative part of the gonad (Fig. 70) with a consequent depression in the development of reproductive cells.

The normal development of the reproductive system in Siberian sturgeon from the Ob River had been well studied (Votinov 1958, 1963). However, these studies do not contain information on whether there were any abnormalities occurring in gameto- and gonadogenesis at the end of the 1950's and beginning of the 1960's. Considering the reproductive performance in the other major Siberian river systems, we undertook a histological investigation of the gonads in a small sample collected in July – August of 1995, 200 km above the Ob Bay (Ruban 1996).

Of the three available females, only the youngest (age 1-2 years, weight 0.2 kg) exhibited normal gonad development, i.e. formation of egg-bearing laminations and migrating oogonia. The second female (age 33 years, body length 165 cm, weight 30 kg) gonads were in maturation stage IV (incomplete). Microscopy showed many abnormalities in gametogenesis: 53% of the oocytes had irregular forms and 87% exhibited stratification of the yolk envelopes and local exfoliate of the jelly envelope (Fig. 71). Furthermore, a large amount of pigment granules were observed among the oocytes indicating complete resorption of some of the oocytes already during trophoplasmatic growth. Thus, this female exhibited two processes simultaneously: growth and development of the majority of oocytes and their partial resorption (Fig. 72). Cavities in the yolk near the vegetative pole, which are not observed

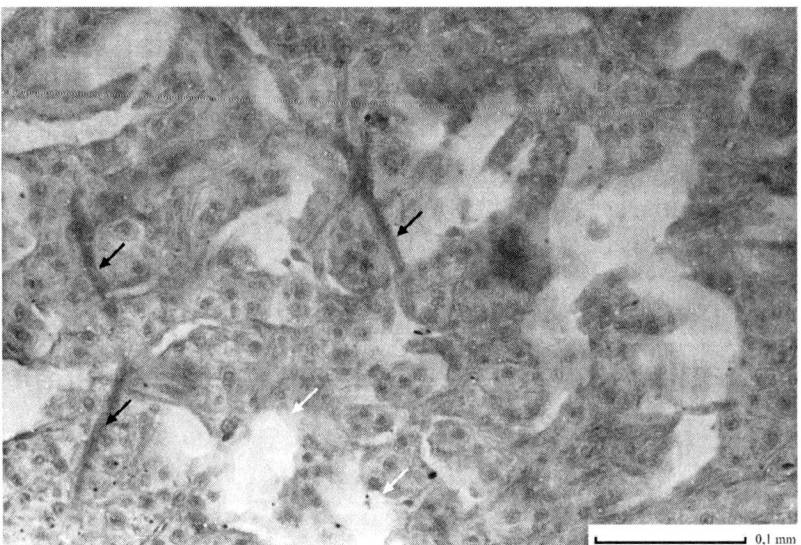

Fig. 70: Anomalies in the microstructure of the gonads of a Siberian sturgeon male from the Yenisei River (II-fatty stage of maturation, total body length 100 cm, body mass 4700 g; age 19 years, magnification: objective 9x ocular 12,5); local disruption of seminal canals and developing cavities between them (indicated by white arrows), abnormal growth of connective tissue in the generative part of the gonads (indicated by black arrow).

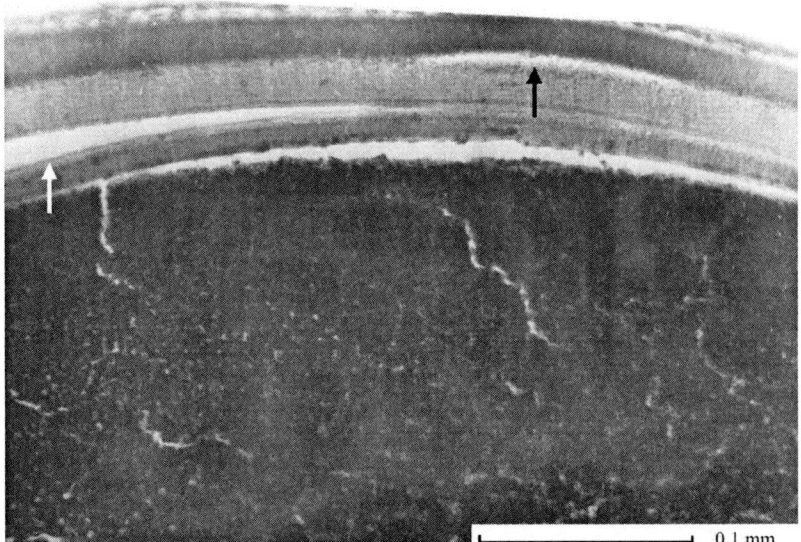

Fig. 71: Disturbances in the microstructure of the gonads of a Siberian sturgeon female from the Ob River (IV stage of maturation, total body length 165 cm; body mass 30 kg, magnification: objective 9x ocular 12,5); fragments of oocytes with local separation of yolk membranes (indicated by white arrow) and separation of the gelatinous membrane (indicated by black arrow);.

under normal development, were observed in 6.7% of the oocytes.

The third female (weight about 30 kg) exhibited a large tumour (15 cm in diameter) (Fig. 73) on the gonads (weight roughly 5 kg). Histological analysis of the gonads showed that the generative tissue was overgrown with adipose tissue and contained enormous fat cells with poorly expressed borders. In the gonads there were large quantities of connective tissue but no oocytes. This individual was sterile (Ruban 1996).

Results of microscopic investigation of the two available males showed that one young individual (body length 57.5 cm, weight 0.5 kg, age 3 years) exhibited normal development of the reproductive cells and gonads. The second mature individual (weight 11.6 kg, age 25 years) with gonads in stage III of maturation exhibited abnormalities in both gametogenesis and in the structure of sexual glands: (a) destruction of the walls of seminal channels with cavities disposed between them and (b) lysis of the spermatozoids near these cavities. In some parts of the gonads, these cavities were rather large.

Although we are aware that this small sample size can hardly be considered representative, the occurrence of Ob River sturgeon exhibiting drastic abnormalities in gameto- and gonadogenesis leading up to total sterility of a female, is at least alarming. In contrast to the Indigirka River population, it appears that the largest disturbances occur in mature fish.

In conclusion, the populations from the Indigirka, Kolyma, and Ob rivers seem to be in the worst condition judging from the pathologies they exhibit. The percentage of females exhibiting abnormalities in these rivers varies between 83 and 100%. Just sturgeon females from these rivers exhibit the highest percentages of oocytes with various abnormalities, causing total sterility in certain cases. In the Lena and Yenisei rivers deviations in gameto- and gonadogenesis is significantly less in both sexes. However, in these cases taking into account a low reproductive potential (resulted by late maturation and long inter-spawning intervals for females), even small pathological changes in the development and functioning of the reproductive system can significantly affect reproductive success and subsequently influence recruitment.

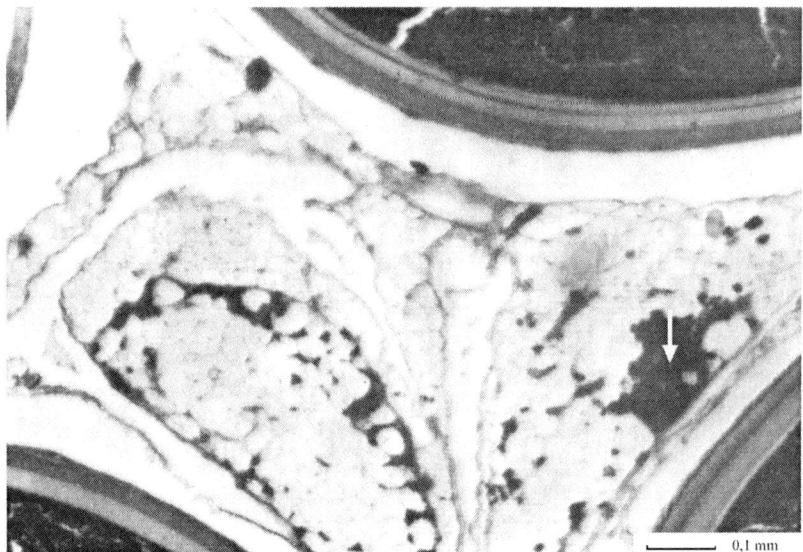

Fig. 72: Disturbances in the microstructure of the gonads of a Siberian sturgeon female from the Ob River (IV incomplete stage of maturation, total body length 165 cm, body mass 30 kg, magnification: objective 3,5x ocular 12,5); accumulations of pigment granules which indicate resorption of some oocytes during trophoplasmic growth (indicated by arrow).

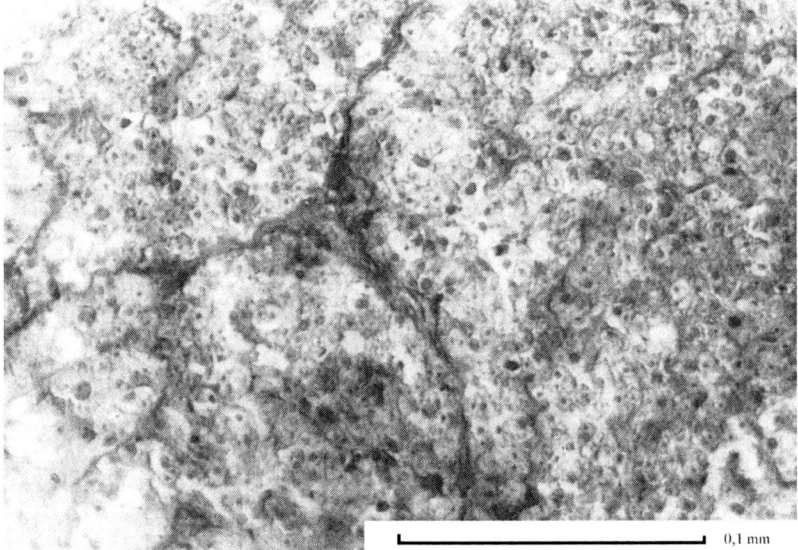

Fig. 73: Disturbances in the development of gonads in a Siberian sturgeon female from the Ob river - a sterile female, body mass 30 kg, microstructure of swelling tissues in the ovary, magnification: objective 20x ocular 12,5).

5.3 A classification of morphological and other abnormalities in sturgeon reproduction

The purpose of this section is to classify the abnormalities the development and functioning of the reproductive system in sturgeon. The classification is based on well-known principles of organizational levels in living systems: cell, tissue, organ, organism, cohort and population, (Seleye 1951, Shmalgauzen 1961, 1982, Ushakov 1963, Naumov 1964, Abramova 1967, Vedenov and Kremyanski 1967, Shvarts 1980, Yablokov and Yusufov 1981).

Our multi-year study on the subject in the Siberian sturgeon from different locations (the Lena, Indigirka, Kolyma, Yenisei rivers, warm-water farming), and for sterlet from the Yenisei River has demonstrated a wide spectrum of abnormalities in gameto- and gonadogenesis (Akimova 1978, 1985a, b, Akimova and Ruban 1992, Akimova et al. 1995a, b, Ruban and Akimova 1991, 1993, Ruban 1996) which we wish now to classify, including analogous data previously obtained on reproductive abnormalities and deviations among acipenserids (Fadeeva 1980, Faleeva 1987, Romanov 1990, Romanav and Altufiev 1990, 1992, Romanov et al. 1990, Zhuravleva et al. 1991, Romanov and Seveleva 1992 and 1993), We propose to consider five levels of reproductive disturbances corresponding to five general levels of biological organization mentioned above: cell level, tissue, organ, organism and populations (Akimova and Ruban 1996).

I. Abnormalities at the Cell Level

Females

A. Pre-vitellogenesis

1) Deformation of a part of oocytes (observed in specimens from all studied populations of sturgeons, excluding Siberian sturgeon from the Lena River and sterlet sturgeon from the Yenisey River);
2) Vacuolization of oocyte karyoplasms (observed in Russian sturgeon from the Volga River, Siberian sturgeon and sterlet sturgeon from the Yenisey River);
3) Abnormalities in cytoplasm structure, its "fragmentation" and uneven staining of the fragments in histological preparations (observed only in sturgeons from Caspian Sea basin);
4) Direct division of oocytes: amitosis (observed in all studied populations of sturgeons, excluding Siberian sturgeon from the Lena River in 1960-1970s and from Indigirka River).

B. Vitellogenesis

1) Deformation of portions of the oocytes (observed in specimen from all studied populations of sturgeons, excluding Siberian sturgeon from the Lena River);
2) Morphological abnormalities of the oocyte's envelopes:
a. Local thinning of the jelly envelope (observed in specimens from all studied populations, excluding Siberian sturgeon from the Lena and Indigirka rivers);
b. Presence of vacuolated bodies of unknown origin within the jelly envelope (observed in females from all studied populations, excluding Siberian sturgeon from the Lena and Indigirka rivers);
c. Exfoliation of the oocyte envelopes (jelly and

yolk envelopes) and uneven staining, indicating biochemical changes within the envelopes (observed in Siberian sturgeon and sterlet sturgeon from the Yenisey River and sturgeons from Volga-Caspian basin);

3) Rupture of the nuclear membrane in immature oocytes (observed in specimen from all studied populations of sturgeons, excluding Siberian sturgeon from the Indigirka and Yenisei rivers and also from the Lena River in 1960-1970s);

4) Formation of cavities filled with a substance of unknown origin beneath the envelopes and among the yolk granules (observed only in sturgeons from the Volga River).

Males

1) Reduction of motility and fertilization ability of sperm (observed in sturgeons from Caspian basin)

II. Abnormalities at the tissue level

Females

1) Asynchronous development of the oocytes during trophoplasmatic growth (observed in Siberian sturgeon from the Indigirka River and from warm water hatchery) including terminal phases of maturation (observed in Siberian sturgeon from the Kolyma river);

2) Partial and mass resorption of the oocytes during cytoplasmatic growth (observed in females from all studied populations of sturgeons);

3) Partial or mass resorption of the oocytes during trophoplasmatic growth (observed in females from all studied populations of sturgeons);

4) Presence of a large number of residual eggs in the gonads after spawning (observed in females from all studied populations of sturgeons excluding Siberian sturgeon from the Kolyma and Yenisei rivers, females of the Siberian sturgeon after spawning in warm water hatchery were not studied);

5) Accumulation of the blood cells in the gonad tissue during cyto- and trophoplasmatic growth (observed in females from all studied populations of sturgeons);

6) Occurrence of gametocytes (gonions and oocytes) in the liver tissues (observed in anadromous sturgeons in Caspian basin during marine and riverine life stages).

Males

1) Local destruction of seminal channels and formation of cavities between them (observed in the Siberian sturgeon from the Kolyma and Yenisei rivers, sterlet from Yenisei river and Russian sturgeon from the Volga river);

2) Large local accumulations of blood cells within the generative part of the gonads (observed in males from all studied populations of sturgeons);

3) Accumulation of adipose tissue between seminal channels (observed in Russian sturgeon from the Volga river);

4) Asynchronous spermatogenesis (observed in Russian sturgeon from the Volga river);

5) Replacement of some parts of testes with liver tissue and vice versa (observed in anadromous sturgeons in Caspian basin during marine and riverine life stages).

III. Abnormalities at the organ level

Females

1) Various forms of hermaphroditism;

2) Discrepancy between the development of the gonads and age, size and weight of the fish (delay in the development of the generative part of the gonads);

3) Inclusions of striated muscles tissue within ovaries;

4) Tumours, cysts, and other neoplasms on the gonad surface.

Anomalies 1, 2, 3, were observed only in sturgeon females from the Caspian Sea basin, 4 – in Caspian sturgeons and also Siberian sturgeon from the Ob river.

Males

1) Presence of connective tissues within the generative part of the gonads (observed in Siberian sturgeon from the Yenisey River and Caspian sturgeons during marine life stage);
2) Appearance of tumours, cysts, and other neoplasms on the surface of the testes (observed in Caspian sturgeons during marine life stage).

IV. Abnormalities at the organism level

1) A decrease of the total and relative fecundity of a female sturgeon (due to the degeneration of a part of oocytes, the large number of eggs retained in gonads after termination of spawning, and the replacement of the generative part of gonads by the connective tissue);
2) Change of the gonadosomatic index (GSI) related to a decrease of the gonad's weight in relation to total weight (due to the degeneration of parts of the oocytes in females and by the damage of seminal channels in males; this is also related to the formation of neoplasms in the gonads and the replacement of their generative parts by connective tissue in both sexes);
3) Change of the gametogenesis rate due to a decrease or increase of duration of cyto- and trophoplasmatic growth periods of oocytes;

V. Abnormalities at the population level

The decline of the reproductive capacity in the entire population due to:

1) Skipping over some spawning periods by the fish with massive resorption of mature gametes; noticeable extension of the spawning interval by several years longer than normal in individual fish. These fish frequently show severe resorption of mature gametes and therefore less females of the population than normally expected participate in the spawning activity each year.
2) Increase of inter-spawning intervals due to mass resorption of gametes;
3) Decrease of population fecundity due to the decrease of individual fecundity, both absolute and relative;
4) Increase in the number of abnormal and unviable offspring caused by the poor condition of gametes.

5.4 Conclusion

Common reproductive abnormalities caused by anthropogenic impact in many fish species (including sturgeons) are: hermaphroditism, degeneration and resorption of gametes, amitosis in oocytes during the cytoplasmatic growth, destruction of the nuclear membrane in immature oocytes, and replacement of seminal channels by connective tissue, accompanied by the formation of multiple cavities (Sakun 1964, Faleeva 1965, Butskaya 1976, Belova *et al.* 1993).

The reproduction of the Volga-Caspian Basin sturgeons is the most widely affected at all of the above identified organisational levels. Despite the fact that reproductive abnormalities in Siberian sturgeon have mainly been found at the cell and tissue levels, eventually they can be manifested at higher levels of organization i.e. organism and population, through changes in individual fecundity and natural reproduction. However, the need for gaining reliable estimates on the significance of such abnormalities on the population level, we are in urgent need of quantifying studies that allow - over time and space – a trend analysis how these classified morphological malformation features are changing in a population.

Currently available classifications of the organization levels in biological systems are to some extent relative and differ in number of separate levels and also in their definitions given by different authors (Shmalgauzen 1961, 1982; Ushakov 1963, Naumov 1964, Abramova 1967, Vedenov and Kremyanski 1967, Shvarts 1980, Yusufov and Yablokov 1981). These differences depend on available information and on the formulation of particular problems. In our case, the classification we used embraces almost all of the available information on reproductive abnormalities in sturgeon related to the development and functioning of the reproductive system. Undoubtedly as new information becomes available, this classification may be modified. Clearly, the properties of a system are more than a simple sum of the properties of its components (Bocharov 1990). That is why reproductive disturbances at a higher level of system organization (e.g. population) cannot be accounted for entirely to the sum of abnormalities at lower levels (cell, tissue, organ and organism) unless these studies are followed by a quantifying assessment at the population level and correlated to reproductive success. Nonetheless, the proposed classification highlights the types of functional units in the reproductive system in sturgeons with the potential to influence the reproduction at population level (e.g. recruitment into the adult stock). As one of such effects of declining level of the reproductive population can be considered the decrease in the absolute individual fecundity (organism level) caused by the replacement of a part of generative gonad tissue by connective or muscle tissues (organ level), by degeneration and resorption of some oocytes during trophoplasmatic growth (tissue level) and by various early pathologies in gametes (cell level). Another sequence of events with significance at population level can be expressed in delayed maturation and an increase in inter-spawning intervals (organism level) caused by the retardation in gonad development (organ level) and the effect may be exaggerated to significance level by massive resorption of oocytes during the cyto- and trophoplasmatic growth, which requires additional time for the resorption and formation of a new oocyte generation (tissue level).

In our opinion, the proposed classification of reproductive disturbances in sturgeon can be a useful tool in standardizing investigational procedures in reproductive success of sturgeons and thereby can greatly assist in clarifying in future studies the causal relationships between abnormalities at different organizational levels and the finite effect on the populations.

6. Stock size of Siberian sturgeon and assessment of it's populations state

Compared to other sturgeon species, Siberian sturgeon harvest has remained relatively modest (Votinov et al. 1975, Gundriser et al. 1983, Yegorov 1988, Dryagin 1949).[1] Maximum harvest was attained in the 1930's peaking at 1,700 tonnes per year. The greatest catches occurred in the Yenisei

---

[1] Data from Nizhnekolymsk regional inspection of Fish protection, CDREN (Central Directorate of Fishery Expertise and Normatives), and the Ministry of Wildlife Protection of the Sakha Republic (Yakutia) are also used is this chapter.

and Ob basins (Fig. 74, 75). Catches in the Lena River have been lower (Fig. 76). The catch in other water bodies was insignificant (Figs. 77, 78, Tab. 53).

For a long time in main sturgeon rivers of Siberia the general tendency of catches was a decline. In 1991 the entire harvest of Siberian sturgeon was 78 tonnes. This dramatic drop in the harvest is most notable in the Ob basin, where the net harvest in 1996 amounted to only 6.7 tonnes. In 1998 the Ob population of the Siberian sturgeon was included in Russian Red Book on endangered species and it's harvest was completely banned. Catches in the Yenisei River before 1998 have likewise been stable and relatively small, ranging from 30 to 48.2 tonnes/year. In 1998 the sturgeon harvest in the Yenisei River was banned and up today it's fishing is permitted only for scientific purposes and for artificial breeding at hatcheries. Sturgeon catches in this river last years are about 10 tonnes per year. The overall decline certainly indicates the dramatic decline in the abundance of sturgeon but does not provide us with any sound measure of the true population size nor on the population dynamics.

In assessing the general and overall population state of the Siberian sturgeon, it must be taken into consideration that a significant part of the habitat is located north of the Arctic Circle and that far northern ecosystems are particularly susceptible to anthropogenic influence (Votinov 1963, Kirillov 1965a,b, 1972, 1982, Dormidontov 1965, Venglinskiy 1974, 1984, Votinov et al. 1975, Tyaptirgyanov 1988). This influence includes harvest (fishing), dam construction, the acquisition of non-metallic resources in the channels of the rivers, pollution from industrial runoff, sewage and other environmental changes, natural and man-made.

Currently, complete data on the total quantity of Siberian sturgeon based on results of direct counts, estimates on natural mortality and fishing mortality and even on fishery statistics do not exist. Therefore, a quantitative assessment or a comparison of state of the separate populations of Siberian sturgeon can only be done using data on catch, changes in the extent of the habitat, and the results of the investigation of the development and functioning of the reproductive system of the sturgeon. The last serve as general indicators of the level of anthropogenic impact and characterizes the weakened potential of natural reproduction and reproductive success of the species.

As can be seen from the results presented above, the types of impacts on separate populations of sturgeon are not same. Of the various Siberian sturgeon habitats, the environmental conditions of the Ob basin, the most developed of the Siberian regions, have been most dramatically transformed. The main factors enacting a negative impact on the Siberian sturgeon in the Ob-Irtysh basin are (a) excessive harvest, (b) dam construction, (c) intensive pollution in the Ob-Irtysh basin due to oil extraction, the influence of the chemical, petrochemical and military industries and the consequent development of river transport, (d) the destruction of spawning grounds due to dam construction, gravel mining and the littering of the rivers due to timber-rafting, (e) illegal fish harvest, exceeding the official limits by a factor of 20 (Ruban 1996). Together, these factors have led to a massive disruption in gameto- and gonadogenesis in the Ob sturgeon, up to total sterility of females, a steep drop in numbers

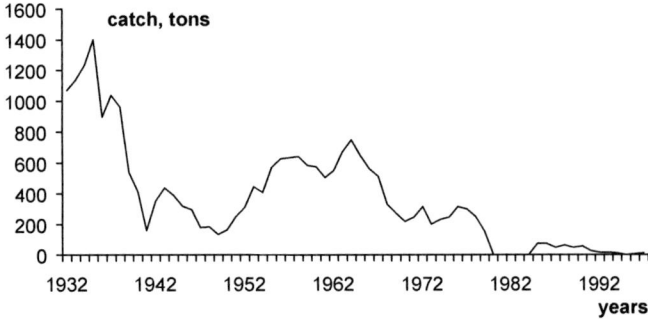

Figure 74: Catches of Siberian sturgeon in the Ob River basin.

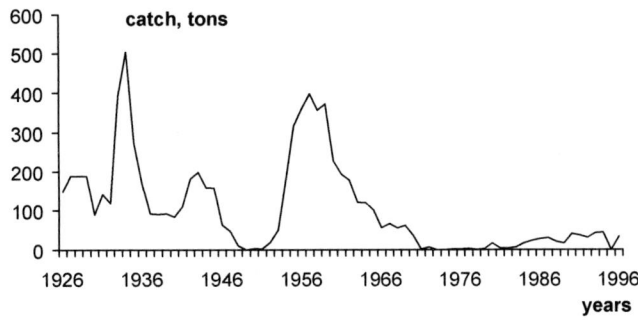

Figure 75: Catches of Siberian sturgeon in the Yenisei River basin.

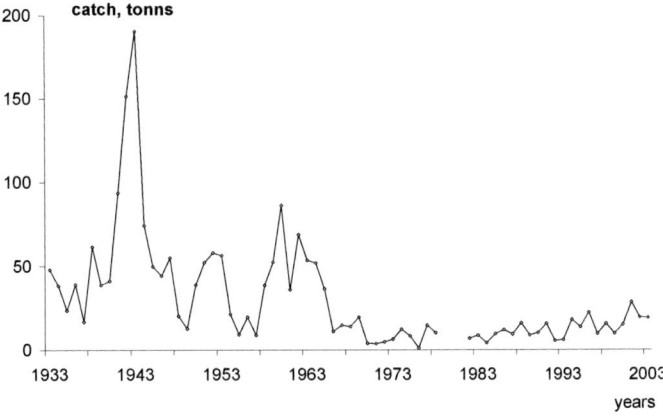

Figure 76: Catches of Siberian sturgeon in the Lena River basin.

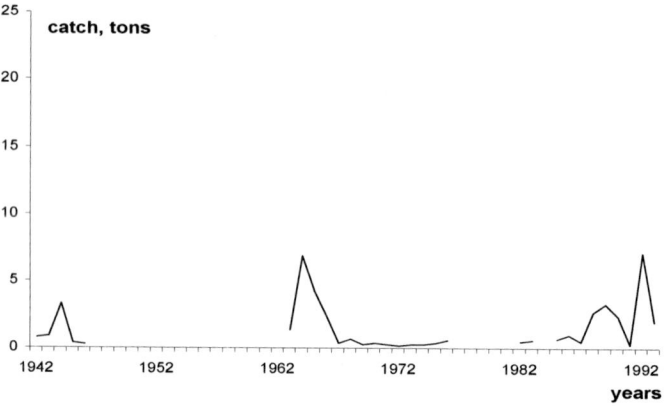

Figure 77: Catches of Siberian sturgeon in the Kolyma River basin.

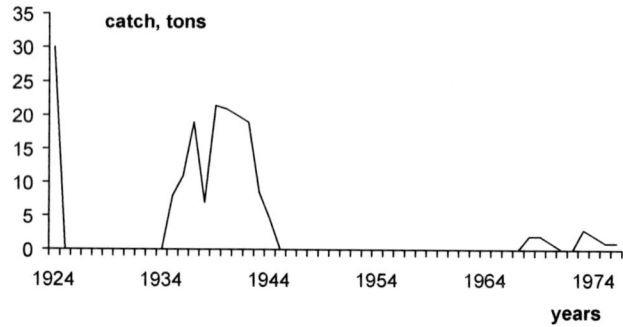

Figure 78: Catches of Siberian sturgeon in Lake Baikal.

Table 53:
Siberian sturgeon annual catches in the Yana and Indigirka rivers, metric tonnes during the period 1942 – 1995 (with reliable records for most of the last decades missing).

Year	River	
	Yana	Indigirka
1942	1.0	0.1
1943	0.4	0.5
1944	1.10	1.3
1945	0.6	0.1
1946	0.3	
1955		16.9
1956		1.6
1961		0.5
1964		0.06
1968		0.01
1993		0.1
1995		0.2

and, consequently, in harvest, and a reduction of the habitat of this population, once the most extensive of the Siberian sturgeon populations. Over 60 years, from 1935-1996, the annual catches of sturgeons in the Ob-Irtysh basin have fallen drastically by an overall factor of 210, from 1,410 tones to just 6.7 tonnes.

The relative influence of each of these factors on the population of Ob sturgeon has varied with time (Matkovsky 1997). From the beginning of 1930's to early 1940's, both mature and young sturgeons were overfished in the Ob delta and the Ob Bay. After the Ob Bay sturgeon fishery was closed, an equally excessive harvest took place in the delta and in the lower reaches until 1969 (Gundriser et al. 1983). Massive fishing of juvenile sturgeon with deep-water seines was practiced from 1941 to 1969 in the middle reaches of the Ob River. In the Tomsk region alone, 120,000 individuals from 18 to 25 cm in length were caught during the hundred day fishing season in 1957, 90% of the total amount of the sturgeon caught here (Gundriser et al. 1983). During 12 years after 1957 damage due to overfishing was further aggravated by deterioration of conditions of reproduction as a result of dam construction. The dam of Novosibirsk HES, have cut off 40% of the spawning grounds and 50% of the wintering grounds for the Ob sturgeon. As a result of damming all of the spawning grounds in the middle and upper Irtysh River were practically eliminated. Damming also have changed the hydrological regimes of the Ob and Irtysh rivers. In this period (late 1950's and early 1960's) the industrial oil extraction in the Ob region had started, leading to elevated pollution levels. The combination of these unfavourable conditions let to dramatic decrease in the natural reproduction of the sturgeon. The count of juveniles migrating downstream near Hanty-Mansiysk city (Ob near mouth of the Nazym River) has shown that the numbers of young caught in one seine per day fell from 400 to 0.3 fish from 1952 to 1963. Later, this number grew somewhat, though in 1971 it has not exceeded on average 25.9 fish/day (Votinov et al. 1975). From the beginning of the 1990's, there has been a further catastrophic fall in the Ob sturgeon population. Reproduction has declined quickly. Mean catch of juveniles in one standard control catch during seaward migration carried out from 1990 to 1995 has fallen from 704 to 80 fish, or by a factor of 8.8 (Krokhalevsky 1997). In 1998, the Ob population of Siberian sturgeon was included in the Russian Red Book on endangered species.

Earlier, the Yenisei population of the Siberian sturgeon was second only to the Ob population. Overfishing in the 1930's and 1950's (Gundriser et al. 1983) was the main cause of its quantitative decline. In the beginning of the 20th century the catch of sturgeon did not exceed 120-150 tonnes per year. In the 1930's, however, the intensity of the harvest grew rapidly and peaked at 393 and 504 tonnes in 1933 and 1934 respectively. Afterwards, the catch fell to 80 tonnes in the 1940's (Fig. 75). During World War II, limitations on fish catches did not exist and sturgeon catches varied from 100 to 200 metric tonnes per year. In 1948-1952, the sturgeon fishery was banned (Gundriser 1983). When it again reopened, the harvest regained high values of 317-398 tonnes per year. For a long time, an intensive harvest of sturgeon took place in the delta and the Yenisei Inlet, the bycatch of juveniles varied from 20% to 65% (Gundriser 1983). As a result of this excessive catch, the stocks of the Yenisei sturgeon were undermined. The generations that experienced heavy harvest twice, as young immature sturgeon in the 1930's and later as mature fish in the 1950's, suffered most. There was once again a complete ban on Yenisei sturgeon fishing from 1971-1990; however, the expected recovery of the population did not occur despite its long-term effect (Podlesniy 1955-1958, Mikhalyov 1991). This is most likely due to the fact that for such a long-lived fish as Siberian sturgeon, even a ban of this duration is not sufficient for reestablishing former numbers because it was of equal or a somewhat shorter period than a single life cycle from hatching to sexual maturation (first spawn) (20 years) (Podlesniy 1955). Another reason for the slow recovery of the population is the decrease of quantity of sexually mature fish from the second half of the 1950's onward, when their annual harvest exceeded 300 tonnes, whereas the catch of mature sturgeon in 1933-1934 never exceeded 140 tonnes (Gundriser et al. 1983). Due to the persistently low stock abundance, the harvest of Siberian sturgeon in the Yenisei River has been banned since 1998.

Dam construction in the Yenisei River has not had as significant an impact on the sturgeon quantity and reproduction as in the Ob-Irtysh basin since all of the significant spawning grounds are located downstream of the Krasnoyarsk HES dam. Upsteram the Krasnoyarsk, sturgeons were rare even before the dams were built. But as a result of their construction sturgeon disappeared in this upper part of the river. The total range of the habitat has been shortened by about 600 km. Last years there were undertaken some attempts to reintroduce youngs of the sturgeon in Sayano-Shushenskoye reservoir by Research Institute of Ecology of Fisheries Water Bodies and Surface Ecosystems (RIEFWBSE), and now young sturgeons again occurs in this reservoir (personal communication by senior researcher of RIEFWBSE V. A. Zadelyenov). The level of gameto- and gonadogenesis abnormalities is not as great in the Yenisei as in the Ob, Indigirka, and Kolyma rivers.

In the second half of the 19th century, the quantity of the Siberian sturgeon population in Lake Baikal was rather high, providing a consistent harvest around 200-300 tonnes per year. However, excessive harvest in the beginning of the 20th century based on catching of spawners during the spawning migration and by the unchecked extermination of immature specimens over the entire habitat, must have led to a rapid fall in the population size and, consequently, expressed itself in reduced harvest. In 1924, the net harvest of sturgeons on two of the most important fishing grounds, the Barguzinskiy and Verkhneudinskiy (Selengiskiy), was a mere 3.87 tonnes (Afanasiev 1997). The ban of Baikal sturgeon fishing introduced from 1930 to 1935 did not result in the recovery of the population. In 1945, the ban was renewed and is still in effect. During the 30-35 years since the renewal of the ban, there has not been an increase in sturgeon stock abundance. From 1985-1988, their numbers were estimated to be about 10,000-18,000 in the Selenga shallows, and 3,000-4,000 fish in the Barguzin Bay. From 1986-1988, only 70-140 adults have entered the Selenga River to spawn (Afanasiev 1997). Due to this extremely low general population size and the low numbers of spawners, the Baikal sturgeon was listed in the Red Book of Russian Federation and classified as a rare and disappearing population (Pavlov et al. 1994).

In the Lena River basin, numbers of sturgeons and corresponding catches have always been significantly lower than in the Ob and Yenisei basins. As for the populations of the Indigirka and Kolyma rivers, the current condition of this (Lena) population can be roughly assessed on the basis of catch dynamics, habitat change and condition of the reproductive system as gauged by the level of abnormalities in the development and functioning of the reproductive system.

The magnitude of the Siberian sturgeon harvest in the Lena River ranges widely though it has, on the whole, fallen during last years (Fig. 76). The harvest peaked in 1941-1945 at 49.3 to 189.9 tonnes. In the first post-war years (1946-1954), catches of Lena sturgeon fell to 12-57.5 tonnes. After a brief period (1955-1957) during which the Lena harvest was rather small 8.1-18.8 tonnes, it grew to rather high level of 35.2-85.7 tons over an extended period (1958-1965). Beginning in 1970, there has been a rapid fall in harvest continuing to the present day. In recent years, the Lena sturgeon harvest has not exceeded 27,8 tonnes and has fallen as low as 0.6 tonnes in some years (1976).

The habitat of the Siberian sturgeon in the Lena basin has been shortened over the past 150 years. The southern border of the range has moved roughly 300 km to the north. The Lena basin was insignificantly affected by dam construction, though the construction of the Viluyskoye reservoir led to the disappearance of the sturgeon from the flooded zone.

Studies of the development and functioning of the reproductive systems of the Lena sturgeon has shown the appearance of different abnormalities and increase of their frequency in recent years. Number and types of abnormalities as well as the ratio of individuals affected by them are similar in the Lena River and Yenisei River populations. These two populations are relatively safe in comparison with those from the rivers Ob, Indigirka and Kolyma. The Indigirka and Kolyma populations are already small, and the high observed level of reproductive abnormalities (Ruban and Akimova 1991, 1993; Ruban 1994, 1996, 1997), affecting 81-100% of the females, justifies a characterization of their status as at high risk.

Taking the existing data on stock abundance, habitat dynamics, natural reproduction rates and the level of reproduction abnormalities of different populations of Siberian sturgeon, it can be stated that the populations of the Ob-Irtysh basin, Lake Baikal, Indigirka and Kolyma are currently in the worst state and at high risk.

7. Adaptive features

The basic theoretical propositions on mechanisms of adaptive radiation, specificity of growth processes, maturation, metabolism, size-age structure, and fish ontogeny developed by G. V. Nikolskii (1974) and N. L. Gerbilskii (1957, 1962, 1965, 1967) were taken into account in the analysis of adaptive features of the Siberian sturgeon.

Gerbilskii (1957, 1962, 1965, 1967) developed a theory on the biological progress of sturgeons based on the ideas of A. N. Severtsov on the biological progress on certain taxonomic groups of animals. The main biological characteristics of such groups are the high number, wide distributional ranges and adaptive radiation. This theory was postulated on the basis of studies on biological features of representatives of the genus Acipenser. The main focus was on anadromous species of the Ponto-Caspian basin. The freshwater species, such as Sterlet sturgeon (Acipenser ruthenus) and Siberian sturgeon were not considered. Gerbilskii (1957, 1962, 1965, 1967, 1972) investigated and in a general outline formulated the system of adaptations providing the increasing degree of euryoeky, intensification of reproduction and reduction of offspring mortality.

The majority of adaptive features inherent in the entire genus *Acipenser* are also characteristic for the Siberian sturgeon living in fresh waters. However, this species exhibits a number of essential differences to the Ponto-Caspian species which should not be reduced only to the distinctions of anadromous and nonmigrating forms. We shall consider them using the scheme of adaptation systems proposed by Gerbilskii (1962, 1967).

The first set of adaptations was named by N. L. Gerbilskii (1962; 1967) as «many-sided ecological fitness» or «increasing of euryoky degree». It includes some interconnected features of species which first of all characterize adaptive radiation within a genus, separate species and populations: amphybionic and holobionic forms, biological groups, seasonal races etc., spatial and temporal differentiation in utilization of foraging resources of water bodies, nature of a migration impulse, in time of spawning and location of spawning sites. Thus, this group of adaptations includes the features of species characterizing first of all their structure.

The analysis of phenotipic diversity and ecological traits of the Siberian sturgeon allowed us to consider it as a monotypic species. The absence of significant differences in morphological characters of the Siberian sturgeon from different river basins, especially among the northern populations, is probably connected with two causes. Firstly, with features of colonization and isolation of separate populations. Acipenserid remnants in Siberia have been found dating back to the Oligocene (37-22 million years) and the Pliocene (5-1.8 million years) (Shtylko 1934, Berg 1962, Gardiner 1984). The glaciation and huge freshwater near-glacial lakes at that time (Baulin 1970) could have had a distributional effect on the range of the Siberian sturgeon (Baulin 1970). Most likely, the complete isolation of some of its populations occupying separate river basins has occurred rather recently. The last, Zyryanian, glaciation terminated about 10,000 years ago (Imbry and Imbry, 1988). Second, the morphological similarity of the northern populations can also be convergential as a result of living under similar climatic conditions. This hypothesis is supported by the phenomenon of parallel latitudinal clinal variation in several morphological characters within large river basins (Ruban 1989, 1992, Ruban and Panaiotidi 1994).

It can be assumed that the Siberian sturgeon as a freshwater species has been formed under the influence of multiple isolation events from the seas caused by glaciation during the quaternary period (Baulin 1970) as well as the lack of sufficiently productive zones in the Arctic Ocean. Consequently, adaptive radiation of Siberian sturgeon was directed not to the formation of distinct ecological forms, typical for anadromous sturgeon (particularly of the Ponto-Caspian basin), but to the formation of a continuous rows of populations, so called "population continuums" of resident forms over a wide range in a single basin. Because of the lack of anadromous forms migrating at various distances in the rivers, such structure allows this species to use efficiently both foraging and spawning grounds of extended river basins, using other mechanisms of spatial and temporal differentiation in spawning.

Because of short hydrological summers in Siberia, the existence of seasonal races of sturgeon, i.e. different groups of sturgeon that reproduce in the same spawning ground at different times is impossible. The occupation and use of distributional range, including spawning habitat, occurs via the formation of population continua along the river, allowing different populations inhabiting different sections of the river to reproduce in the optimal time without performing energetically expensive migrations. It gives certain advantages under conditions of the limited food supply in northern rivers. This population structure is typical over the most part of the Siberian sturgeon's distributional range. There are only two exceptions caused by specific ecological conditions of water bodies and these are identified below.

The population inhabiting the Ob-Irtysh basin has a more complex structure than the populations of other Siberian rivers. Besides the resident forms, there is there is also so-called «anadromous form» performing potamadromous migrations. The principal differences of this form from anadromous sturgeons inhabiting the Ponto-Caspian basin were described above. It is obvious that the existence of this form is determined by the unique features of the Ob basin, the annual winter hypoxia, which compels all fish to abandon the lower reaches of the Ob and Irtysh rivers (Votinov 1963). In the absence of this event, the population structure of the Ob sturgeon would most likely be similar to that of the populations in other large river basins. Other authors (Malyutin 1980) share this opinion. In contrast to anadromous species of the Ponto-Caspian basin, the Siberian sturgeons migration in the Ob River basin triggered by the oxygen saturation of water.

A separation between foraging and spawning grounds is characteristic for the Siberian sturgeon inhabiting basin of Lake Baikal. On the one hand, food supply in the spawning rivers (Selenga, Bagruzin, Upper Angara) is rather scarce. On the other hand, Lake Baikal does not provide appropriate spawning grounds. Thus, the sturgeon is compelled to perform extended potamadromous migrations from the feeding grounds of Lake Baikal to spawning grounds in the rivers.

The population structure, including multiple age cohorts, is typical of all sturgeon species, and this holds for the Siberian sturgeon as well. This structure provides certain advantages under unstable and changing environmental conditions, since low numbers or even total collapse of a single generation does not have a too great impact on the population as a whole (Nikolskii 1974). The features of the age and sexual structure of the Siberian sturgeon populations such as liability of life span ratio of males and females depending on environment conditions may be interpreted as adaptive. As was shown above, the life span of males and females is similar in populations with higher food supply and high growth rate. In populations living under worse foraging conditions, whose growth rate is much lower, the life span of males is almost twice less than that of females. For the majority of sturgeon species, it is typical the numerical prevalence of males at spawning grounds, but in populations of the Siberian sturgeon this phenomenon is less expressed, and more often the sex ratio at spawning grounds is close to 1:1. This allows the species to maintain high reproduction levels under low food supply conditions.

The first group of adaptations according to Gerbilskii (1962, 1967), providing high degrees of euryoeky, also includes feeding patterns. All acipenserids are to some extent omnivorous. However, in the Siberian sturgeon this property is more expressed than in the various species of the Caspian sturgeons. In the overwhelming majority of cases the Siberian sturgeon is a non-selective benthivore. The diet composition reflects the composition of benthos at foraging grounds (Romanova 1984). Moreover, only the Siberian sturgeon has populations (Lake Baikal) that are able to switch almost entirely from benthic feeding to piscivory at a certain age. This allows the Baikal sturgeon to attain even greater growth rates than the fastest growing riverine sturgeon (the Ob population) inhabiting the most favourable thermal regime. A high level of omnivory is further supported by the above described observations for the Indigirka sturgeon foraging by drown rodents. Thus, intraspecific differentiation on the type of feeding in various populations of the Siberian sturgeon from primary benthivorous to mainly carnivorous feeding, reaches a level close to distinctions between representatives of the genus *Acipenser* and *Huso* in the Caspian Sea.

As an adaptation, related to scarce of Siberian sturgeon foraging source, can be considered the fact that most populations of the sturgeon feed at the main part of the range through the winter at the water temperature close to 0°C and that populations in the more severe north-eastern parts of the range feed through the spawning period as well.

Following N. L. Gerbilskii (1967), two other groups of adaptations include the ecological features of sturgeons promoting intensification of their reproduction and reduction of offspring mortality. It is obvious that these groups of adaptations are interconnected and their division is rather conventional. It concerns such features of sturgeons as wide limits of spawning temperatures and embryonic tolerance temperatures. The analysis of our materials and of data of other authors shows that the Siberian sturgeon does not only spawn at lower average temperatures than sturgeons of southern water bodies, but it can develop under much wider upper and lower limits of temperature. The optimal temperature for embryonic development of the Siberian sturgeon is about 11.4-14.9°C (Nikolskaya and Sytina 1974), whereas for stellate sturgeon from the Don River and Russian sturgeon from the Black Sea, the temperature ranges are 17-24°C and 15-21°C, respectively (Detlaff and Ginzburg 1954). The threshold temperatures of embryonic development are 8-25°C for Siberian sturgeon, 12-25°C for stellate sturgeon in the Volga River, 12-22°C for stellate sturgeon migrating in the Don River in spring, 10-20°C for Russian sturgeon migrating in the Don River in spring, and 14-23°C for Russian sturgeon migrating to the Kura River in autumn (Detlaf and Ginzburg 1954).

Some physiological features of the Siberian sturgeon connected with maturation can be considered as adaptations promoting intensification of its reproduction. In general, acipenserids are characterized by late maturation. The Siberian sturgeon likewise matures at a relatively late age, though, as shown above, it displays considerably more plasticity in the relationship between growth and age to sexual maturity in dependence on food supply. The size of maturing females in high-growth rate populations (Ob, Baikal) is roughly 1.5 times greater than in the slow-growing populations of the Lena, Indigirka and Kolyma river systems. The age at sexual maturity for the fast-growing populations is likewise greater. These differences are highly significant. Females of the Baikal sturgeon, whose growth rate is high, spawn for the first time at the age which is almost 2 times more (20-22 years), than that in females of the sturgeon from the Lena (11-13 years). Observations from warm-water hatcheries cultivating Lena sturgeon have provided evidence of the relationship of size and age at sexual maturity to the availability of food. In an environment with elevated temperatures and abundant food, the Lena sturgeon, typically one of the slowest growing of the Siberian sturgeon in it's natural habitat, results in a lowered age at maturity (factor of about 1.5) and an increase in length (factor also 1.5). The linear size of maturing females reared under such conditions attains the size of the first spawning females from the Ob population whose growth rate is high. Sexual maturation in the majority of fish species is connected with reaching a certain size characteristic of these species independently on age (Nikolskii 1974). However, at the boundaries of the habitat range there are deviations in age and size at maturity from the given rule, expressed in the change of the relationship between generative metabolism and somatic growth (Shatunovsky 1986).

The significant differences in the level of generative metabolism were found among populations of the Siberian sturgeon with different growth rates. The greatest values of the gonadosomatic index (up to 66,8 %) were observed in females from the Lena River, whose growth rate was the lowest. This parameter in females from the Ob River with high growth rate was much lower (up to 33 %). The females from the Lena River at the repeated sexual cycles have the greatest values of relative fecundity and high content of fat in gonads at IV stage of maturity in comparison with other riverine populations (Akimova 1985a, b).

Shatunovskii (1986) discovered some metabolic adaptations of fish in the northern limits of their geographical range. At these parts of the range fish usually have a low efficiency of somatic growth, and often elevated levels of energetic metabolism and generative tissue synthesis. In populations of the Siberian sturgeon living under most severe conditions in their marginal habitat (Lena and Indigirka rivers) we also find the smallest growth rates and a significant increase in generative metabolism coupled with a decrease of somatic growth efficiency. It resulted in maturation at earlier age, smaller size, increase of relative fecundity, and relative gonad weight of females. The increase of generative metabolism level was accompanied by an increase in duration of interspawning intervals of females, which were not less than 5 years in this part of the range. These features can be interpreted as adaptive and directed to the maintenance of high numbers under conditions of low food supply.

According to the classification of Gerbilskii (1967), such features of Siberian sturgeon development as accelerated postembryonic development of its larvae in comparison with southern species and the absence of a pelagic feeding period can be attributed to adaptations minimizing offspring mortality. Siberian sturgeon prelarvae convert to a near-bottom lifestyle almost immediately after yolk absorption (Malyutin 1980).

These features are based on general relationships between the rate and efficiency of yolk utilization, growth, and development of larvae in relation to temperature during embryonic and postembryonic periods. A decrease in temperature results in hatching of larger prelarva with a greater amount of residual yolk and also greater protein increase of prelarvae during the period from hatching to the beginning of a mixed diet (Novikov 1991). The environmental stressors other than temperature cause similar responses in terms of length of incubation time and size at hatching, such as oxygen concentration (Brett 1976), the exposure to heavy metal concentrations (Rosenthal and Sperling, 1974).

Siberian sturgeon developing at lower temperatures than the Ponto-Caspian sturgeons, has larger sizes of hatched prelarva than the Russian sturgeon (10.5 mm and 9.4-9.6 mm respectively) (Chusovitina 1963, Detlaf et al. 1981). The weight of mature eggs in Russian sturgeon varies from 15.1 to 26.0 mg (Trusov 1964, Barnnikova 1970, Amirkhanov 1972). Siberian sturgeon ripe eggs are a little smaller ranging from 10.8 to 25 mg (Sokolov 1965, Sokolov and Malyutin 1977, Akimova 1978). These data show that the efficiency of yolk utilization in the Siberian sturgeon at lower temperatures is higher than in Russian sturgeon. It is known that a decrease in temperatures during the period from hatching to the stage of mixed feeding (simultaneous yolk and exogenous feeding) results in an increase in prelarvae protein ascertion (Novikov 1991). The latter probably also explains the larger size of the Siberian sturgeon larvae at the stage of transition to active feeding in comparison with the Russian sturgeon larvae developing at higher temperature. Differences in the size of larvae between Siberian sturgeon and Russian sturgeon are about 12% at the end of the yolk sac stage. Most likely, the smaller amount of fat content in mid-intestine cells and in the spiral fold of the Siberian sturgeon prelarvae during the transition to mixed feeding can be explained by more complete yolk utilization for protein growth in the Siberian sturgeon compared to the Russian sturgeon. Thus, the resistance to starvation is lower in Siberian sturgeon prelarvae than in Russian sturgeon. Russian sturgeon larvae undertake an extended downstream migration, whereas the transition to external (exogenous) feeding in Siberian sturgeon larvae takes place near the spawning grounds (Bogdanova 1972).

Thus, it seems to be most probable that the Siberian sturgeon developing at low temperature utilizes the yolk more efficiently than Ponto-Caspian sturgeons which results in larger larvae at the stage of transition to the active feeding, and near-bottom life style.

These features during early ontogeny in Siberian sturgeon can be considered to be adaptations to northern water bodies with short vegetation periods and low planktonic biomass in the riverbeds.

8. Perspectives on conservation and restoration of sturgeon populations

The extent of anthropogenic influence on different Siberian sturgeon populations is not equal and is related to the intensity of fishing, the scale of hydro-construction, the level of pollution and human transformations of the water bodies. Therefore, the degree of decline of natural reproduction and subsequent population size, are different. Appropriate conservation measures for the protection of Siberian sturgeon populations must take into account the types of anthropogenic impacts imposed on separate populations. The populations in Lake Baikal, the Ob-Irtysh basin, and the Indigirka and Kolyma rivers, for example, are in the worst condition. However, the individual factors that have brought these populations to this poor state are rather different. The dramatic state of the Baikal sturgeon population is due almost exclusively to overfishing, whereas the threatened condition of the Kolyma and Indigirka sturgeon, having always been low in quantity, is due to the massive abnormalities of the reproductive system caused by rivers pollution. The decrease of the Ob population size is related to a combination of these factors. At various points in history, this population was threatened by: (1) overfishing and a high harvest of immature fish; (2) construction of dams, which cut off the migratory populations from the upper Ob and Irtysh and their significant spawning grounds; (3) the unprecedented pollution of the Ob basin leading to massive abnormalities in the development and functioning of the reproductive systems (up to total sterility of females). Based on these facts it can be concluded that measures designed to conserve and restore populations of the Siberian sturgeon must be specific for each of the populations.

The monotypicity of Siberian sturgeon and its high ecological plasticity does not justify a conservation strategy based on introducing fish from other populations since the uniqueness of the species requires the conservation of its genetic structure. Conservation measures for the recovery of Siberian sturgeon populations must provide conservation (and restoration) of natural genetic and phenetic diversity, i.e. of all the population groups. Thus, the conservation of Siberian sturgeon must be organized around two main directions: 1) the development and realization of a complete set of measures directed to conservation of sturgeon natural populations, including strict regulation of fisheries, protection, a restoration of spawning grounds and artificial reproduction, the lowering of pollution levels in all water bodies, and 2) the establishment of "living collections" of acipenserids, including the different populations of the Siberian sturgeon in adequately designed facilities (Artyukhin and Romanov 1997, Barannikova 1993, 1995).

8.1 Conservation measures for Siberian sturgeon

From the data presented above, it can be concluded that of all the major Siberian sturgeon populations, only the Yenisei River population is in a relatively favourable state. However, even in the Yenisei, a declining population has justified the closing of the fishery.

In the Ob River prior to 1998, sturgeon were allowed to be harvested only as bycatch during whitefish fisheries. However, due to the extremely low numbers of this population, the catastrophic drop in natural reproduction, and an increase in poaching which amounts to over 20 times greater than the officially permitted catch (Ruban 1996), the Ob population had recently to be listed in the Russian Red Book.

The average annual harvest of the Lena sturgeon over a ten-year period from 1993 to 2002 was 15.1 tonnes with a maximum of 27.8 tonnes. It seems a limit need to be established not exceeding 15 tonnes per year on the Lena sturgeon, including licensed catch by fisherman.

Taking the reproductive state of the already small Indigirka and Kolyma Siberian sturgeon populations into account (Ruban and Akimova 1991, 1993), their harvest should be totally banned. Furthermore, the spawning grounds in the Indigirka River (310 and 330 km from the mouth of the river) and in the Kolyma River (adjacent to the mouth of the Ozhigin River and downstream to Srendnekolymsk) must be protected.

Likewise, in other Siberian rivers where the populations of Siberian sturgeon have always been low, such as the Pyasina, Khatanga, Anabar, Olenyok and Yana rivers, their harvest should be banned. The potential catch would not yield an important increase in the total catch, but risks the destruction of these small populations. The possibility for restoration, however, is in practice unrealistic.

8.2 Measures for the restoration and artificial reproduction of Siberian sturgeon

For populations such as in the Ob River and Lake Baikal, once providing the highest catches and currently under threat of extinction, it is necessary to establish large-scale measures for the development of their artificial reproduction with the aim of future restocking.

It has been estimated that the production of roughly 13 million fingerlings per year seems to be necessary for the recovery of the Ob population. This requires the construction of three new sturgeon hatcheries (at Hanty-Mansiysk, Kolpashevo, and Novosibirsk) and the reconstruction of the existing Abalak sturgeon hatchery (Krokhalevskii 1997). These activities are required to maintain the bulk of the Ob sturgeon below the Novosibirsk HES dam. In the upper Ob and in the Novosibirsk reservoir, a local resident population of sturgeon has formed a self-sustaining population. In recent years, there has been an increase of juvenile fish in this population around the mouth of the Verkhnyaya Inya river. This increase is estimated with a factor of 9.5-15.1, which originate from individuals isolated in the upper half of the Ob River upstream of the dam (Solovov 1997). Spawning grounds in the upper Ob River after the damming of the river at Novosibirsk were reduced from 8,200 ha to 850 ha. Currently, spawning grounds in the Biya and Aley and in the lower reaches of the Katun have been destroyed. Consequently, a fish hatchery around the lower Charish must be established to compensate for this loss of spawning habitat (Solovov 1997).

For a long time, the VostSibRibNII-Project (Eastern Siberian Fisheries Research and Designing Institute) undertook research on artificial cultivation of the Baikal sturgeon. Biotechnology of hatching and rearing of young fish was developed (Afanasieva 1984) and in recent years, young Baikal sturgeons have been released from the Selenga hatchery. Undoubtedly, artificial cultivation must be broadened and it must continue not only up to the time of renewal of natural reproduction in rivers but up to the time of population sizes that indicate a restoration level comparable to the population sizes at the end of the 1930's, when annual catches amounted to 20 tonnes.

The transformation of the Angara River into a series of reservoirs has drastically changed the environmental conditions for sturgeons. For most of its length, the Angara River has become unsuitable for Siberian sturgeon habitation or reproduction. However, introductions of young Baikal sturgeon into the Bratskoye reservoir indicate that artificially hatched sturgeon might be able to be maintained in the reservoir (Afanasyev and Polyakov 1986). It is necessary to continue hatchery work with the use of adults concentrating in the Belaya River mouth as spawners. These spawners might likewise be usable for restoration of the population in Lake Baikal. Before the filling of the Boguchanskoye reservoir with water, sturgeon entered the Irikneyeva and Chuna rivers to spawn, both of which entirely lost their significance in this function as a result of timber rafting. Therefore, to establish sturgeon populations in the lower reaches of the Angara River, there must be a clean up of habitats damaged by the timber industry remains followed by a mass introduction of juvenile Siberian sturgeon into the reservoir. It is suggested that the Yenisei sturgeon may be the best source suited for this purpose, since they are closest to the populations that once inhabited the lower reaches of the Angara River.

8.3 Conservation of the Siberian sturgeon gene pool

The other main strategic option for the conservation of Siberian sturgeon populations will be an expansion of the work on building captive brood stocks of acipenserids from various riverine origin. The main goal of these "live collections" will be to conserve the gene pool and aid in re-establishing declining and gradually disappearing natural populations. The practical realization of these measures can progress by including separate Siberian sturgeon populations in regionally localized commercial aquaculture (the wide use of Lena sturgeon in aquaculture is a good example). There is also a need to establish specialized fish hatcheries (such as of the Mozhaisk Experimental Fish Hatchery) specifically dedicated to the conservation of the living gene pool of acipenserids (Artyukhin and Romanov 1997). Economically, the first method (inclusion of different populations into commercial sturgeon aquaculture) appears more practical but – because of the economic pressures – inherits the potential risk of inadequate handling and maintenance because of the need of high density culture and the need for fast growth. Since commercial aquaculture of Siberian sturgeon will mainly have to be based on warmwater aquaculture, there will be a need to establish cold water aquaculture for the restoration of natural populations and this will have to be done in parallel to commercial production in order to produce juveniles fit for release in natural water bodies.

Over the past several decades, the work on the establishment of brood stocks of separate populations of Siberian sturgeon has expanded. Currently, the fish hatchery division of the Konakovo Industrial Experimental Branch of All Russian Research Institute of Freshwater Fisheries, the aquaculture department of the Krasnodar power station, the Volgorechensk hatchery the warm-water hatchery at the Gusinoozersk power station have developed breeding pools of Lena and Baikal sturgeon. The Aqua Company (Yuzhno-Uralsk, Chelyabinsk region) has developed a brood stock of Ob sturgeon. Currently, it is necessary to expand these projects by including other populations of sturgeon, especially from the Indigirka, Kolyma and Yenisei rivers. The creation and maintenance of breeding pools in fish hatcheries can guarantee the possibility of restoration of populations. Particularly, the recovery of the Baikal population began with the creation of a breeding pool, which consequently allowed for the possibility of artificial reproduction.

The necessary preconditions for the conservation are known today and realistic recovery strategies for the Siberian sturgeon populations are ready for implementation. The main measures for conservation include not only (a) a total ban on catches of threatened populations (Pyasina, Khatanga, Anabar, Olennyok, Yana, Indigirka and Kolyma), but also (b) the strict enforcement of such bans. Furthermore, the anti-poaching enforcement needs to be strengthened especially in the Ob and Yenisei river catchments, while also the pollution load to these rivers must be reduced. This holds especially for the Ob, Indigirka and Kolyma rivers. As an additional point sturgeon specimens from the Yenisei, Indigirka and Kolyma populations should be included into the existing living collections of acipenserids, while the massive stock enhancement programs for the Ob and Baikal sturgeon must be continued.

Measures for the recovery of Siberian sturgeon populations must focus first on the Ob and Baikal populations. For this, the release of roughly 13 million fingerlings of Ob sturgeon a year is necessary, the construction of three new hatcheries in the middle and lower Ob (Hanty-Mansiysk, Kolpashevo, Novosibirsk), the reconstruction of the Abalak sturgeon hatchery, the construction of a sturgeon hatchery near the lower reaches of the Charysh river, the restoration of spawning grounds in the Biya and Aley rivers and in the lower reaches of the Katun River, the expansion of artificial reproduction and an increase of Baikal sturgeon releases. In order to restore Siberian sturgeon populations in the Angara basin, fish must be introduced into the Bratskskoye and Boguchanskoye reservoirs, and the Irikneyeva and Chuna rivers must be cleaned up of timber industry remains.

References

Publications in Russian

Abramova, N. T. 1967. On the Definition of the "Organization Level" Concept. Pages 185-201 in Strukturnye urovni biosistem (The Levels of Biosystems Structure). Conf. Proc. Moscow: Akad. Nauk SSSR";

Abuel'ez, A. S. 1990 a. The Combined Effect of Phenol and Propanid on Juvenile Russian Sturgeon. Pages 62-67 in Ekologo-fiziologicheskie i toksikologicheskie aspekty i metody rybokhozyaistvennykh issledovanii (Ecological, Physiological and Toxicological Aspects and Methods of Fishery Studies). Moscow: All-Union Research Institute for Sea Fisheries and Oceanography (VNIRO).

Abuel 'ez, A. S. 1990 b. Combined Effect of Phenol and Metaphos on the Juveniles of Russian Sturgeon. Pages 67-73 in Ekologo-fiziologicheskie i toksikologicheskie aspekty i metody rybokhozyaistvennykh issledovanii (Ecological, Physiological and Toxicological Aspects and Methods of Fishery Studies). Moscow: All-Union Research Institute for Sea Fisheries and Oceanography (VNIRO).

Afanasiev, G.A. 1997. Condition of Baikal sturgeon stock. Page 104 in Proceedings of the First Ichthyological Congress of Russia. Astrakhan. Izdatedetstvo of All-Union Research Institute for Sea Fisheries and Oceanography (VNIRO), Moscow

Afanasieva, V. G. 1977. On the artificial stock enhancement of Baikal sturgeon. Pages 139-141 in Fish and and fish husbandry in Eastern Siberia vol. 1. Ulan-Ude.

Afanasieva, V.G. 1984. Artificial cultivation of Baikal sturgeon. Pages 26-27 in Sturgeon husbandry in the USSR from Proceedings the all-Soviet conference, December 11-14 1984. Astrakhan

Afanasieva, V.G. and O.A. Polyakov. 1986. The introduction of Baikal sturgeon into the Bratskoye reservoir. Pages 6-7 in Ichthyology, hydrobiology, hydrochemistry, entomology and parasitology. Proceedings of the IX All-soviet symposium "Biological Problems of the North". Akademia Nauk, Yakutsk.

Akimova, N.V. 1978. Gametogenesis, functioning of the sexual glands in Siberian sturgeon (Acipenser baeri Brandt) in the Lena River and their link with metabolites. Pages 43-55 in B.V. Koshelev, editor. Ekologomorfoligicheskie i ekologo fiziologicheskie issledovaniya razvitiya ryb (Ecomorphological and ecophysiological studies of fish development). Nauka, Moscow.

Akimova,N.V.1981. Maturation and sexual cycles of sturgeons (with special reference to Siberian sturgeon from the Lena River). Pages 48-57 in B.V. Koshelev and M.V. Gulidov, editors. Reproduction and development of fishes (methodic text-book). Moskva, Nauka.

Akimova, N. V. 1985a. Gametogenesis and reproductive cycles of the Siberian sturgeon under natural and experimental conditions. Pages 111-122 in Osobennosti reproduktivnikh tsiklov u ryb v vodoemakh raznykh shirot USSR (Characteristics of reproductive cycles in fish at different latitudes of the USSR). Moscow, Nauka

Akimova, N. V. 1985b. Gametogenesis and reproduction in Siberian sturgeon. PhD dissertation. Institute of Morphological

Evolution and Animal Ecology, USSR Academy of Sciences. Moscow. 243 pp.

Akimova, N. V. 1988. Energetic metabolism in Siberian sturgeon under different environmental condidtions. Pages 4-5 in Proc. of the All-Soviet Conference on Ecological Energetics of Animals. Puschino.

Akimova, N.V. 1996. Assessment of the reproductive system.Pages 285-290 in Ekologicheskaia bezopasnost 'i ustoichivoe razvitiye Samarskoi oblasti (Ecological safety and the sustainable development of the Samara Region), No. 3.Toliatti.

Akimova, N. V., G.I. Ruban. 1992. The condition of the reproductive system of the Siberian sturgeon, *Acipenser baeri,* as a bioindicator. Voprosy Ikhtiologii (Issues in Ichtyology). 32:102-109 (English translation: J. Ichthyol. 33: 15-24).

Akimova, N. V., G. I. Ruban. 1995. The response of the reproductive system in acipenserids to anthropogenic influence as a factor in population dynamics. Pages 491-500 in Proceedings of the Russian Conference on Population Ecology: Structure and Dynamics (15-18 November 1994, Puschino), vol. II. Russian Academy of Sciences, Moscow.

Akimova, N.V., A.I. Panaiotidi, and G.I. Ruban. 1995a. Disturbances in Development and Functioning of Reproductive System in Sturgeon (*Acipenseridae*) from the Yenisei River, Voprosy Ichtiologii (Issues in Ichtyology), 35: 236-246

Akimova, N. V., G. I. Ruban and Yu. V. Mikhalyev. 1995b. Analysis of the state of reproductive system of the Siberian sturgeon in Central Siberia. Pages 93-98 in *I. A.* Shilov, editor. Northern Ecosystems: Structure, Adaptations, Stability. Izdatelstvo Moskovskogo Gosudarstvennogo Universiteta (Moscow State University Publishers), Moscow

Akimova, N. V., G. I. Ruban. 1996. A classification of reproductive disturbances in sturgeons (Acipenseridae) caused by anthropogenic impact. Voprosy Ichtiologii (Issues in Ichtyology). 36: 65-80 (English translation: Journal of Ichthyology. 36: 61-76).

Akimova, N.V., L.1. Sokolov, I. Smoljanov and V.S. Malyutin. 1980. Comparative analyses of growth and gametogenesis of the Lena River Siberian sturgeon in natural and experimental conditions. Pages 167- 176 in E. I. Vorobyeva, editor. Intraspecific variability in animals onthogenesis. Nauka, Moscow.

Alabaster, J. and R. Lloyd. 1984. Water Quality Criteria for Freshwater Fish, translated from English. Legkaya i Pishchevaya Promyshlennost, Moscow

Alekin, O. A. 1970. The principles of hydrochemistry. Leningrad.: Hydrometeoizdat. 296 p.

Aleksandrov, A.V., Yu.S. Kotov, and F.S. Bilalov. 1988. Distribution of Heavy Metals in the Main Elements of Water Ecosystems. Page 4, Proceedings of V Vsesoyuznaya konferentsiya po vodnoi toksikologii (V All-Union Conf. on Water Toxicology), Moscow.

Alekseev, F. E. 1969. Hermaphroditism and Regulation of Sexual Structure in the Population of *Pagellus acarne* (Risso) in Some Areas of the Nothwest Coast of Africa, Tr. Atlant. NII Rybn. Khoz. Okeanograf (Studies of the Atlantic Research Institute of Fisheries and Oceanagraphy) 22:21-31.

Amirkhanov, M.I. 1972. The state of sturgeon gonads during the spawning migration in the Terek River. Trudy Tsentralnogo Nauchno-Issledovatelskogo Instituta Osetrovogo Rybnogo Khozyaistva 4: 26- 29.

Andreev, V.L. 1980. Klassifrkatsionnye postroeniya v ekologii i sistematike (Classification constructions in Ecology and Systematics), Nauka, Moscow.

Andrienko, E.K., V.P. Krokhalevski, V.A. Slepokurov and V.I. Uvarova. 1997. Results of the ecological monitoring of the Ob River. Pages 102-103 in Proceedings of the First Ichthyological Congress of Russia. Astrakhan. Izdatedetstvo VNIRO, Moscow.

Andriyashev, A.P. 1954. Fishes of the northern seas of the USSR. Izdatelstvo Akademii Nauk USSR, Moscow-Leningrad. 566 pp.

Andronov, A.E. 1983. A Study of the Intrapopulation Differentiation in Stellate Sturgeon in the Lower Volga on the Basis of Gonad State. Pages 61-72 in Biologicheskie osnovy osetrovodstva (The Biological Bases of Sturgeon Farming). Nauka, Moscow.

Anokhin, Yu.L. and F.V. Muzyka. 1988. Chronic Intoxication in Fish Affected by Combined Pollution. Pages 4-5 in V vsesoyuznaya konferentsiya po vodnoi toksikologii (V All-Union Conf. on Water Toxicology), Moscow.

Anpilova, V.M. 1965. Sex Redifferentiation in Baunt Whitefish *Coregonus lavaretus baunti* Muchomedyarov, affected by Ecological Conditions, Voprosy Ichtiologii (Issues in Ichtyology). 5: 207-209.

Antonov, V.S. 1956. Spring and fall water-level patterns in the Lena River. Trudy Arkticheskogo Nauchno-Issledovatelnogo Instituta. 6: 18-61

Antonov, V.S. 1960. The Lena Deta (brief hydrological notes). Trudy Okeanografitcheskoi komissii AN SSSR (Proceedings of Comission on Oceanography, Academy of Sciences USSR). 6:25-34.

Argentov, O.A. 1860. Fish of the Kolyma River system, adjacent lakes and the Arctic Ocean. Pages 352-368 in Akklimatizatsiya vol. 1 No. 8 Isdatelstelstvo Komiteta Aklimatizatsii.

Artyukhin, E.N. and A. G. Romanov. 1997. The Mozhaisk esperimental fish hatchery's mission to conserve a living gene fund of acipenserids. Page 8 in Proceedings of the First Ichthyological Congress of Russia. Astrakhan. Izdatedetstvo VNIRO, Moscow.

Askhaev, M. G and E. S. Gomenyuk. 1966. Characteristics of the development of the ichthyofauna of the Irkutsk reservoir. Conference on the biological productivity of Siberian water bodies, Irkutsk. 55-56

Averina, I. M., V. G. Agapitov, N. A. Dorogina, N. G. Zhadrinskaya, Yu. A. Kruchinin, G. L. Rutilevskiy, R. K. Sisko, I. V. Semenov. 1962. Northern Yakutia (physical geography). Trudy Arkticheskogo i Antarkticheskogo nauchno-issledovatelskogo instituta Glavsevmorputy MMF SSSR (Transactions of Arctic and Antarctic Research Institute, General Department of North Sea Way, Ministry of Navy USSR). Leningrad. 236: 280 pp.

Averintsev, S. V. 1930. Fish Reserves of Yakutian Water Bodies and Perspectives for their Expoitation. Bulletin Rybnovo Khozaistva (Bulletin of Commercial Fisheries). 4:19-20.

Averintsev, S.V. 1933a. An Expedition of the Fisheries Station in the lower reaches and delta of the Lena River in the Summer of 1930. Tr. Yakut. nauch. rybkhoz. stantsii (Transactions of the Research Fisheries Station of Yakutia). Leningrad. Isdatelstvo Vsesoyuznogo Nauchno-Issledovatelskogo Instituta Ozernogo i Rechnogo Rybnogo Khozyaistva. II: 7-56

Averintsev, S. V. 1933b. Materials on the biologial statistics of commercial fish in the lower Lena River. Tr. Yakut. nauch. rybkhoz. stantsii

(Transactions of the Research Fisheries Station of Yakutia). Leningrad. Isdatelstvo Vsesoyuznogo Nauchno-Issledovatelskogo Instituta Ozernogo i Rechnogo Rybnogo Khozyaistva. II:87-148.

Averintsev, S. V. 1933c. The status of fisheries in the lower reaches and delta of the Lena River: current condition and ways of development. Tr. Yakut. nauch. rybkhoz. stantsii (Transactions of the Research Fisheries Station of Yakutia). Leningrad. Isdatelstvo Vsesoyuznogo Nauchno-Issledovatelskogo Instituta Ozernogo i Rechnogo Rybnogo Khozyaistva. II: 209-258.

Averintsev, S. V. 1933d. Exploratory investigations of fish reserves in the middle Lena. Tr. Yakut. nauch. rybkhoz. stantsii (Transactions of the Research Fisheries Station of Yakutia). Leningrad. Isdatelstvo Vsesoyuznogo Nauchno-Issledovatelskogo Instituta Ozernogo i Rechnogo Rybnogo Khozyaistva. II: 259-268.

Barannikova, I.A. 1970. New data on the reaction of sturgeon populations on disturbances of migration and reproduction conditions. Trudy Tsentralnogo Nauchno Issledovatelskogo Instituta Osetrovogo Rybnogo Khozyaistva 11: 12 - 19.

Barannikova, I.A. 1975. Functional Foundations of Fish Migrations. Nauka, Leningrad. 210 pp.

Barannikova, I.A. and T.A. Fadeyeva. 1982. Gonadotropic function of stellate sturgeon *Acipenser stellatus* Pallas hypophysis in the marine stage of life, Voprosy Ikhthyologii (Issues in Ichtyology). 22: 460-465.

Baranov, F.I. 1918. On the biological bases of fisheries. Izv. Otd. rybovodstva i nauch.-promysl. issled. 1: 84-128

Baulin, V.V. 1970. The history of the underground glaciacion at West Siberia in connection with the Arctic basin transgression. Pages 404-409 in A.I. Tolmachev, editor. The Arctic Ocean and its seaboard. Leningrad.

Belova, N.V., Verigin, B.V, Emelyanova, N.G. et al. 1993. Radiobiological Analysis of Silver Carp *Hypophthalmichthys molitrix* in the Cooling Reservoir of Chernobyl Power Plant Following the Disaster; 1: State of the Reproductive System in Fish That Survived the Disaster. Voprosy Ikhtiologii (Issues in Ichtyology). 33:814-828.

Belykh, F. 1. 1940. Lake Lama and its fishery use. Trudy Nauchno-Issledovatelskogo Instituta Polyarnogo Zemledeliya. Zhivotnovodstva i Promyslovogo Khozyaistva, Seriya Promyslovoe Khozyaistvo 11: 73-100.

Berdichevskii, L.S., Malyutin, V.S., Smolyanov, I.I., and L.I. Sokolov. 1983. The results of aquaculture and acclimatization researches of the Siberian sturgeon. Pages 259- 269 in I.A. Barannikova and L.S. Berdichevskii, editors. Biologicheskiye osnovy osetrovodstva. Nauka, Moscow.

Berg, L.S. 1900. Fish of the Baikal. Yezhegodnik Zool. Muz. Imp. Akad. Nauk (Annual of the Zoological Museum of the Imperial Academy of Sciences) 5: 326-372.

Berg, L.S. 1905. Fish of Turkestan. Nauchnye Rezultaty Aralskkoi Ekspeditsii.: Izvestiya Russkogo geogrtcheskogo obschestva (Scientific results of the Aral expedition: Transactions of Russian Geographical Society). St. Petersburg. 4:261 pp.

Berg, L.S. 1908a. A List of Kolyma Fishes. Yezhegodnik Zool. Muz. Imp. Akad. Nauk (Annual of the Zoological Museum of the Imperial Academy of Sciences) 13: 70-107.

Berg, L.S. 1908b. A List of Fishes in the Ob basin. Yezhegodnik Zool. Muz. Imp. Akad. Nauk (Annual of the Zoological Museum of the Imperial Academy of Sciences) 13: 221-228.

Berg, L.S. 1911. Fauna Rossii i sopredel'nykh stran. T1: Ryby (Fauna of Russia and Adjacent Countries, vol. 1: Fishes). St. Petersburg. 337 pp.

Berg, L. S. 1916. The freshwater fishes of the Russian empire. Moscow. 563 pp.

Berg, L. S. 1923. The freshwater fishes of Russia. Gosizdat, Moscow. 535 pp.

Berg, L. S. 1926. Fishes of the Khatanga Basin. Materialy Komissii AN SSSR po Izucheniyu Yakutskoi ASSR 2:1-22.

Berg, L.S. 1928. On the evolution of the northern elements in the fauna of the Caspian Sea. Dokl. Akad. Nauk. SSSR, series A, 7: 107-112.

Berg, L. S. 1932. The freshwater fishes of the USSR and adjacent countries, Vol. 1, Part 1. Akademia Nauk SSSR, Moscow & Leningrad (English translation published by Israel Program for Scientific Translations, Jerusalem. 505 pp.)

Berg, L. S. 1949. The freshwater fishes of the USSR and adjacent countries, Vol. 1-3 Akademia Nauk SSSR, Moscow & Leningrad. 1382 pp.

Berg, L.S. 1962. Fishes of the Amur River Basin. Pages 320-360 in G.V. Nikolskii and D.V. Obrutchev, editors. L.S. Berg - Selected transactions. Moskva-Leningrad. V. 5.

Blagovidova, L.A. 1976. Condition of the zoobenthos of reservoir in the second decade of their existence. Pages 83-98 in Biologichiskii rezhim i rybkhozaistvennoye ispol'zovanie Novosibisrskogo Vodokhranilischa. Zap.-Sib. knizhnoye izd., Novosibirsk.

Bocharov, L.N. 1990. Sistemnyi analiz v kratkosrochnom rybopromyslovom prognozirovanii (System Analysis in the Short Term Fishery Forecasts). Nauka, Leningrad. 208 pp.

Bogan, F. E. 1939. Contribution to the biology of the Siberian sturgeon *(Acipenser baeri* Brandt) of the Irtysh River basin. Uchenye Zapiski Permskogo Gosudarstvennogo Universiteta 3:145-163.

Bogdanov, V.D. 1983. Hatching and migration of *Coregonidae* larvae in the Ural tributaries of the Lower Ob. Pages 55-79 in Biologia i ekologia gidrobiontov Nizhnei Obi. UNTs AN SSSR, Sverdlovsk.

Bogdanova, L.S. 1972. The ecological plasticity of sturgeon larvae and youngs. Pages 244 - 250 in Yu.Yu. Marti and I.A. Barannikova, editors. Osetrovye i problemy osetrovogo khozyaistva. Moskva. Pischevaya promyshlennost.

Bolkvadze, L.D. and M.A. Gogotisgvili. 1988. Pathological Changes in the Organs of Pickarel under Intoxication with the SN-79 Compound. Pages 98-103 in Vodnaya toksikologiya I optirnizatsiya bioproduktsionnykh prorsessov v akvakul'ture (Water Toxicology and Optimization of Bioproduction Processes in Aquaculture). MRKh VNIRO Moscow."

Borisov, PG. 1926. The Results of Ichthyological and Fishery Studies in the Lena River, Dokl. Akad. Nauk SSSR, Series A:161-163.

Borisov, P.G. 1927. Essay on fisheries in the Yakut republic. Izd. AN SSSR, Leningrad. 21 pp.

Borisov, P G. 1928a. Fishes of the Lena River. Trudy Komissii ANSSSR po Izucheniyu Yakutskoi Respubliki. Leningrad. 9: 1-181.

Borisov, PG. 1928b. The Modern State of the Fishing Industry in the Lower Reaches of the Lena River and Ways of Its Development,

Materialy Komissii AN SSSR po Izucheniyu Yakutskoi Respubliki, No. 28. 32 pp.

Borisov, P.G. 1929. Preliminary data on the fisheries in the lower Kolyma, Materialy Komissii AN SSSR po Izucheniyu Yakutskoi Respubliki, No. 28.

Braginskii, L.P. and M.I. Kuz'menko. 1988. Ecological and Toxicological Aspects of Design and Exploitation of Nuclear Power Plants, Page 11 *in:* V Vsesoyuznaya konferentsiya po vodnoi toksikologii (V All-Union Conf. on Water Toxicology). Izd. VNIRO, Moscow.

Braginskii, L.P., F.Ya. Komarovskii, E.P. Shcherban', P.N. Linnik, L.F. Osipov. 1989. Ecological and Toxicological Situation in the Water (General Principles of Assessment and Forecast). Gidrobiologicheskii Zhurnal 25: 91-101.

Brodski, V.Ya. 1964. Direct Division of the Nucleus, Uspekhi Sovrem. Biol. 58: 367-394.

Brodski, V.Ya. 1966. Trofika kletki (Cell Alimentation). Nauka, Moscow. 356 pp.

Brodski, V.Ya. and Uryvaeva. 1981. IX, Kletochnaya poliploidiya. Proliferatsiya i differentsirovka (Cell Polyploidy. Proliferation and Differentiation). Nauka, Moscow.

Burlayeva, V.B. 1973. Ichthyofauna of Lower Angara River and influence of the construction of Bogutchanskaya HES on it. Pages 22-23 *in* Vodoyemy Sibiri I perspektivy ikh rybokhozyzaystvennogo ispolzovaniya (Water bodies of Siberia and perspectives of their fisheries use). Tomsk.

Burmakin, E. V. 1941. Some little commercial and noncommercial fishes of the Gyda Bay system. Trudy Nauchno-Issledovatelskogo Instituta Polyarnogo Zemledeliya, Zhivotnovodstva I Promyslovogo Khozyaistva, Seriya Promyslovoe Khozyaistvo 15: 149-158.

Burmakin, E.V. and P.V. Tyurin. 1959. On Biological Classification of Fishes, Voprosy Ikhtiologii (Issues in Ichtyology). 13:19-25

Burtsev I. A. 1962. On the reproductive ability of a hybrid of sturgeon and sterlet. Doklady Academii nauk SSSR 144: 1377-1379.

Burtsev I. A. 1967. Some data on gametogenesis of hybrides of sturgeons. Pages 252-257 *in* Trudy Tzentralnogo nautshno-issledovatelskogo instituta osetrovogo rybnogo khosjaistva. Astrakhan.

Butskaya, N.A. 1976. On Mass Intersexuality in Ruffe *Acerina cernua* (L.) from the Eastern Part of the Bay of Finland. Voprosy Ikhtiologii (Issues in Ichtyology). 16:812-821.

Butskaya, N.A. 1980. The Role of Temperature and Photoperiod in the Sexual Cycle of the Ruffe *Acerina cernua* (L.) (*Perciformes, Percidae*). Voprosy Ikhtiologii (Issues in Ichtyology). 20:849-858.

Chalikov, B.G. 1930. Notes on feeding of *Acipenser baeri* Brandt near Tobolsk city. Materialy po izutcheniyu Sibiri. 1:52-53.

Cherfas, N.B. 1962. The Deterioration of the Carp Gonads Caused by Radiation. Voprosy Ikhtiologii (Issues in Ichtyology). 2:104-115.

Chernovskii, A.A. 1949. Opredelitel Lichinok Komarov Semeistva *Tendipedidae* (Guide to Identification of Mosquite Larvae *Tendipedidae*) AN SSSR, Moscow. 185 pp.

Chernysheva, V.M. 1960. The Response of Fish Gonads from Changes in Ecological Conditions. Pages 181-182 *in* 3 Vsesoyuzn. soveshchanie embriologov (3 All-Union Conf. of Embryologists), Moscow.

Chmilevskii, D.A. and Ivoilov, A.A. 1993. An Attempt at Suppressing Gonad Development in the Females of Tilapia *Oreochromis mossambicus* using Radiation to Stimulate their Growth. Voprosy Ikhtiologii (Issues in Ichtyology). 33(5):732-735.

Chmilevskii, D.A. and N.I. Zakharova. 1987. The Development of Gonads in Rainbow Trout *Salmo gairdneri* Following Radiation Impact; Report II: Irradiation of Fish 5.5 Months After Hatching. Pages 30-39 *in* :Voprosy iskusstvennogo razvedeniya ryb (Problems of Fish artificial breeding), Vsesoyuznyi Nauchno-Issledovatelskyi Institut Ozernogo i Rechnogo Rybnogo Khozyaistva, Leningrad.

Chupretov, V. M. and V. A. Slepokurov. 1979. On winter distribution of the Siberian sturgeon in the Ob and Taz bays. Pages 270-271 *in* V. 1. Lukyanenko (ed.) Sturgeon Fishery in Inland Water Bodies of the USSR, Abstracts of Papers at the 2nd All-Union Conference, Astrakhan.

Chuprov, S.M. and A.A. Vischegorodtzev. 1997. Organization and experience of ichthyological monitoring in the Krasnoyarsk reservoir. Page 180 *in:*Tez. Dokl. 1 Konf. Ikhtiol. Ross. Astrakhan. Moscow.

Chusovitina, L.S. 1963. Postembryonal development of the Siberian sturgeon (*Acipenser baeri* Brandt). Trudy Ob-Tazovskogo otdeleniya Gosudarstvennogo nauchno-issledovatelskogo instituta ozernogo i rechnogo rybnogo khozyaistva (GosNIIORKh) III:103 -114.

Dashi-Dorzhi Anudarin. 1955. Materials on the ichtyofauna of the upper Selenga and Amur rivers within the territory of Mongolia. Zool. zhurnal 34: 570-577.

Davletyarova, R.A., N.A. Kanieva and V.N. Kirillov. 1989. A Characteristic of Some Biochemical Indexes and the Level of Concentration of Chloro-organic Pesticides and Heavy Metals in Russian Sturgeon under Conditions of Increasing Pollution. Pages 75-76 *in* Vsesoyuzn. soveshchanie: Osetrovoe khozyaistvo vodoemov SSSR (All-Union Conf. on the Sturgeon Farming in the USSR Water Reservoirs), Astrakhan.

Davydov, L.K. 1955. Gidrodrafia SSSR (Hydrography of the USSR). Izd. LGU, Leningrad.

Dedyu, I.I., V.I. Ashevskii, L.P. Rogoshevskii, et al. 1988. The Relationship of Biological Variables of Predictive Ecotoxicological Monitoring on Various Abiotic Factors. Pages 30-31 *in* V Vsesoyuzn. konferentsiya po vodnoi toksikologii (V All-Union Conf. on Water Toxicology), Moscow.

Deryugin K.M. 1898. A Journey Into the Valley of the Middle and Lower Ob and the Regional Fauna. Tr. Imperator. Sankt Peterburgsk. obsch. estestvoispytateley (Works of the Imperial Sain Petersburg Society of Natural Explorers) XXIX: 47-140.

Detlaf, T.A., and A.S. Ginzburg. 1954. The embryonal development of sturgeons (Stellate sturgeon, Russian sturgeon, Giant sturgeon) in relation with problems of their breeding. (Zarodyshevoe razvitiye osetrovykh ryb (sevryugi, osetra i belugi) v svyazi s voprosami ikh razvedeniya). Izdatelstvo Akademii Nauk SSSR, Moscow.

Detlaf, T.A., Ginzburg, A.S., and O.I. Shmalgauzen. 1981. The development of sturgeons. Eggs maturation, fertilization, development of embryos and prelarvae. (Razvitie osetrovykh ryb. Sozrevanie yaits, oplodotvorenie, razvitie zarodyshei i predlichinok). Nauka, Moscow.

Dmitriev, V.I. 1941. Fish and fishing in the lower Yenisei River. Pages 7-36 *in* Rybi i Rybniy promysel v nizoviakh reki Yeniseya, v reke Khatange i v Anadyrskom limane. Tr. NII polyarn. zemledeliya, zhivotonovstva i promyslovogo khozaistvo.

Dokholyan, V.K., G.S. Shleifer, T.P. Akhmedova and A.K. Magomedov. 1980. The Influence of Dissolved Oil Products on the Life Functions in Some Fish from the Caspian Sea. Voprosy Ikhtiologii (Issues in Ichtyology). 20: 733-738.

Domanitskii, A.P., R.G. Dubrovina and A.I. Isaeva. 1971. Rivers and lakes of the Soviet Union. Gidrometeoizdat, Leningrad. 104 pp.

Dormidontov, A. S. 1963. Fishery utilization of the Lena River sturgeon. Pages 182-187 in Sturgeon Fishery in the Water Bodies of the USSR, Izdatelstvo Akademii Nauk SSSR, Moscow.

Dormidontov, A. S. 1965. The regulation of the harvest of semi-anadromous fish in the water bodies of Yakutia. Pages 77-81 in Priroda Yakutii i ee Okhrana (The Nature of Yakutia and its Protection). Yakutsk branch of Izvestiya Vsesoyuzniyi Nauchno-Issledovatelskyo Institut Ozernogo i Rechnogo Rybnogo Khozyaistva GOSNIORKh, Yakutsk.

Dormidontov, A. S. and M.P. Sofronov. 1976. The biology of Lower Lena sturgeon. Pages 233-28 in Prirodniye resursy Yakutii, ikh ispolzovaniye i okhrana (The nature resources of Yakutia, their use and protection). Izd. Komisii po okhrany prirody Yakutii Ya.F. SSSR, Yakutsk.

Doronina N.A. 1956. Some hydrographic characteristics of the Lena River. Tr. Arkt. i Antark. Instituta 204: 5-17.

Dryagin, P. A. 1933. Fish resources of Yakutiya. Trudy Soveta po Izucheniyu Proisvoditelnykh Sil 5: 3-94.

Dryagin, P. A. 1947. Sturgeon catches in the water bodies of Siberia. Rybnoe Khozyaistvo 1: 34-38

Dryagin. P. A. 1948a. On some morphological and biological distinctions between sturgeon occurring in the rivers of Yakutiya and the Siberian sturgeon, *Acipenser baerii* Brandt. Zoologicheskii Zhurnal 27: 525-534.

Dryagin, P. A. 1948b. Commercial fishes of the Ob-Irtysh River Basin. Izvestiya Vsesouznogo Instituta Ozernogo i Rechnogo Rybnogo Khozyaistva 25(2): 3-104.

Dryagin, P. A.1949. Biology of the Siberian sturgeon, its reserves and rational utilization. Izvestiya Vsesouznogo Instituta Ozernogo i Rechnogo Rybnogo Khozyaistva 29:3-51.

Dybovskii, B. 1876. Fish of the Baikal water system. Izv. Sib. otd. Imperator. Russk. Geografich. obsch. (Report of the Siberian division of the Imperial Russian Geographical Society) 7: 11-25.

Dyuzhikov, A.T. and E.V. Serebryakova. 1964. Some aspects of the ecology and duration of the reproductive cycle in the sturgeon of the Volga River. Tr. Vses. NII morsk. pibno. khozva i okeanogr. 56: 105-115.

Dzhavadova, L.A. 1992. Influence of Environmental Pollution on the Dynamics of Physiological Functions in Juvenile Sturgeon of Different Ages. Page 19 in Abstract Dissertation, Cand. Sc. (Biol.), Baku.

Egelskii, E.I. 1967. Survival and growth of young Siberian sturgeon (Lena and Baikal populations) in cultivation in pools in Baltic states. Trudy Tsentralnogo Nauchno Issledovatelskogo Instituta Osetrovogo Rybnogo Khozaistva I: 286-290.

Egelskii, E.I. 1970. Growth of young Siberian sturgeon in natural and artificial environments. Trudy Tsentralnogo Nauchno Issledovatelskogo Instituta Osetrovogo Rybnogo Khozaistva II: 191-196.

Ezhegodnik kachestva poverkhnostnikh vod i effektivnosti prevedenykh vodookhrannykh meropriyatii po territorii deyatelnosti Kolymskovo territorialnogo upravleniya po gidrometeorologii Goskomgidrometa za 1989 g. (Annual report on the quality of surface waters and the effectiveness of environmental protection measures in the Kolyma territory of the Goskomgidromet jurisdiction in 1989). 1990b. Magadan. 142 pp.

Ezhegodnik kachestva poverkhnostnikh vod i effektivnosti prevedenykh vodookhrannykh meropriyatii po territorii deyatelnosti Yakutskovo UGKS Goskomgidrometa za 1984 g. (Annual report on the quality of surface waters and the effectiveness of environmental protection measures in the Yakutiya territory of the Goskomgidrometa jurisdiction in 1984). 1985. 161 pp.

Ezhegodnik kachestva poverkhnostnikh vod i effektivnosti prevedenykh vodookhrannykh meropriyatii po territorii deyatelnosti Yakutskovo UGKS Goskomgidrometa za 1985 g. (Annual report on the quality of surface waters and the effectiveness of environmental protection measures in the Yakutiya territory of the Goskomgidrometa jurisdiction in 1985). 1986. Yakutsk. 168 pp.

Ezhegodnik kachestva poverkhnostnikh vod SSSR 1989 (Annual report on the quality of surface waters in the USSR 1989). 1990. VNIIGMI-MTsD, Obninsk. 492 pp.

Ezhegodnik kachestva poverkhnostnikh vod SSSR 1990 (Annual report on the quality of surface waters in the USSR 1990). 1991. VNIIGMI-MTsD, Obninsk. 466 pp.

Faleeva, T.I. 1965. Analysis of ovocyte atresia in fish with regard to its adaptive significance. Voprosy Ikhtiologii (Issues in Ichtyology) 5(3): 455-470.

Faleeva T. I. 1979. The comparative and experimental analysis of anomalies in oogenesis of fishes. Ph.D. Thesis. Leningrad. 268 pp.

Fadeeva, T.A. 1980. A State of Gonads and Size-Age Composition in the Starred Sturgeon Acipenserstellatus Pallas Males from the Caspian Sea During their Marine Life. Voprosy Ikhtiologii (Issues in Ichtyology) 20(6):833-843.

Faleeva, T.I. 1987. The Disturbance of Oocyte Maturation in Stellate Sturgeon in Artificial Cultivation. Voprosy iskustvennogo razvedeniya ryb (Problems of Fish Farming), Leningrad: Izvestiya Vsesoyuznogo Nauchno-Issledovatelskogo Instituta Ozernogo i Rechnogo Rybnogo Khozyaistva. 259:(121-133).

Figurin, A.E. 1823. Notes of a Medico-Surgeon on Various Objects of Natural History and Physics Studied in Ust'ansk and the Adjoining Region in 1822. From G. Spassky (editor) Sibirskiy Vestnik (Siberian Investigator). Saint Petersburg. Book 20-21: 185-213. Book 22: 215-235. Book 23-24: 235-248.

Filenko, O.F. Vodnaya toksikologiya (Water Tixicology). Mosk. Gos. Univ., Chernogolovka. 156 pp.

Gasanov, M.V. 1983. The purification of run-off from oil-extracting and refining industries with the use of microorganisms. Pages 34-36 in Gidrobiologicheskie i ikhtiologicheskie issledovanie w Azerbaijane (Hydrobiological and ichthyological researches in Azerbaijan). Elm, Baku.

Gedenshtrom M.M. 1823. Description of the Coastline of the Arctic Ocean from the Mouth of tha Yana to Rock of Baranova. From G. Spassky (editor) Sibirskiy Vestnik (Siberian Investigator). Saint Petersburg. Book 7: 1-12, 8: 13-26, 9: 27-42.

Geraskin, P.P. 1989. Disturbance of Metabolism in Russian Sturgeon under Current Conditions of the Volgo-Caspian Basin. Pages 60-62 in Vsesoyuzn. Soveshchanie: Osetrovoe khozyaistvo vodoemov SSSR, (All-Union Conf. on the Sturgeons Farming in the USSR Water Reservoirs), Astrakhan.

Geraskin, P.P., N.V. Bal', and E.A. Mishin. 1989. Fractional Protein Composition in the Oocytes of Russian Sturgeon and the Changes under Current Conditions in the Volgo-Caspian Basin. Pages 62-64 in Vsesoyuzn. Soveshchanie: Osetrovoe khozyaistvo vodoemov SSSR, (All-Union Conf. on the Sturgeons Farming in the USSR Water Reservoirs), Astrakhan.

Gerbilskii, N.L. 1957. The directions of development of intraspecific biological differentiation, types of anadromous migrants, and the problem of migration impulse in sturgeons. Uchenye zapiski Leningradskogo Gosudarstvennogo Universiteta. Seria biologicheskikh nauk 228: 11 - 32.

Gerbilskii, N.L. 1962. The theory of sturgeon's biological progress, and its use in practice of sturgeon fishery. Uchenye zapiski Leningradskogo Gosudarstvennogo Universiteta. Seriya biologicheskikh nauk 311: 5- 18.

Gerbilskii, N.L.1965. The theory of biological progress, and its use in fishery. Pages 77-84 in G.V. Nikolskii, editor. Theoretical basis of fishery. Moskva, Nauka.

Gerbilskii, N.L. 1967. The study of functional bases of intraspecific evolution in relation with problems of fish quantity and range in fishery.
Vestnik Leningradskogo Gosudarstvennogo Universiteta 3:5-21.

Ginzburg, A.S. and T.A. Detlaf. 1975. The sturgeon *Acipenser gueldenstaedti*. Pages 217-263 in Obyekti biologii razvitiya. Nauka, Moscow.

Golovinskaya, K.A. and D.D. Romashov. 1958. The Influence of Radiation on the Development and Reproduction of Fish, Voprosy Ikhtiologii (Issues in Ichtyology) 11: 16-38.

Golyshkina R.A. 1967. *Ephemeroptera* of the Angara River and the Irkutskoye reservoir. Izv. Biol.-geografich. nauchnogo instituta gos. univ. Irkutsk. XX: 34-64.

Gorbacheva, L.T. and O.A. Vorobieva. 1979. Reproduction of the Great Sturgeon in the Don River, Pages 63-64 in 1 Vsesoyuzn. soveshchanie: Osetrovoe khozyaistvo vnutrennikh vodoemov SSSR (II All-Union Conf. on the Sturgeon Management in the USSR Inland Waters).

Gorkin, I.N. 1990 Ecological and Physiological Aspects of Bioconcentration of Microelements by the Water Organisms in Natural Environment. Pages 20-34 in Ekologo-fiziologicheskie i toksikologicheskie aspekty i metody rybokhozyaistvennykh issledovanii (Ecological and Physiological Aspect and Methods of the Fishery Research). VNIRO, Moscow.

Gosudarstveniy Doklad o Sostoyanii Okryzhayuschei Sredy Rosssiyskoi Federatzii v 1991 Godu (Government Report on the Condition of the Environment in the Russian Federation in 1991). 1992. Izd. Ministry of Environmentlal Protection and Natural Resoruces of the Russian Federation, Moscow. 80 pp.

Greze, V.N. 1953. Assessment of the biologic productivity of the Yenisei River. Tr. Barabinsk.

References

otd. Vses. NII. ozern. i rechn. ribn. khozaist. VI: 103-105.

Greze, V.N. 1957. Feeding resources of the fish in the Yenisei River and their consumption. Izvestiya Vsesoyuznogo Nauchno-Issledovatelskogo Instituta Ozernogo i Rechnogo Rybnogo Khozyaistva. XLI: 234.

Guide to Freshwater Invertebrates of Europe Part of the USSR. 1977. Gidrometeoizdat. Moscow. 510 pp.

Gundrizer, A. N., A. G. Yegorov, V. G. Afanaseva, S. A. Enshina, Yu. V. Mikhalev, R. I. Setsko & A. A. Khakimullin. 1983. Perspectives of reproduction of Siberian sturgeons. pp. 241-253. *In:* I.A. Barannikova & L. S. Berditchevsky (ed.) Biological Foundations of Sturgeon Management, Nauka, Moscow.

Gusev, A.G. 1971. Biological Basis of Standardization in Conservation of Fishery Reservoirs from Contamination. Pages 29-42 *in* Kriterii toksichnosti i printsipy metodik po vodnoi toksikologii (A Criterion of Toxicity and Principles of the Methods on Water Toxicology), Vsesoyuznyi Nauchno-Issledovatelskyi Institut Ozernogo i Rechnogo Rybnogo Khozyaistva, Leningrad.

Hu Tsi-Tsai, 1957. Differences in development and growth of juveline carp from eggs spawned by one female. Page 67 *in* Vtoroye Sov. Embriologov SSSR. MGU, Moscow.

Hydrochemical Bulletin. 1977. No. 4. Izd. Yakutsk UGKS Goskomgidrometa, Yakutsk. 14 pp.

Hydrochemical Bulletin. 1978. No. 4. Izd. Yakutsk UGKS Goskomgidrometa, Yakutsk. 13 pp.

Hydrochemical Bulletin. 1979. No. 4. Izd. Yakutsk UGKS Goskomgidrometa, Yakutsk. 15 pp.

Hydrochemical Bulletin. 1983a. No. 2. Izd. Yakutsk UGKS Goskomgidrometa, Yakutsk. 14 pp.

Hydrochemical Bulletin. 1983b. No. 3. Izd. Yakutsk UGKS Goskomgidrometa, Yakutsk. 17 pp.

Hydrochemical Bulletin. 1983c. No. 4. Izd. Yakutsk UGKS Goskomgidrometa, Yakutsk. 19 pp.

Imbry, D., and K. Imbry. 1988. Tainy lednikovykh epokh. Poltora veka v poiskakh razgadki (Mysteries of glacial epochs. One and a half century in search of a solution). Progress, Moscow. 263 pp.

International Code of Zoological Nomenclature, Univ. of California: Berkeley, 1985. Translated under the title *Mezhdunarodnyi kodeks zoologicheskoi nomenklatury,* Leningrad: Nauka, 1988. 204 pp.

Iogansen, B.G. Studies on the geography and genesis of Siberian ichthyofauna. Pages 43-60 *in* Ekologo-geograficheskiy ocherk ryb basseina reki Obi (Ecological and geographical description of the fish in the Ob River basin). Uch. zap. Tonsk. Gos. Uni., Tomsk.

Iokhelson, V.I. 1898. Some data on the fish of the Kolyma district. Zemlevedenye III: 75-90.

Isaev A.I. and E.I.Karpova. 1989. Fisheries in reservoirs. Agronomizdat, Moscow. 255pp.

Isachenko, V.L. 1912. Fishes of Turukhanskii Krai Found in the Yenisei and Yenisei Gulf. Materialy po issledovaniyu reky Yeniseia v rybopromyslovom otnoshenii. Krasnoyarsk. No. IV. 111 pp.

Isachenko, V.L. 1916. On the problem of fish diet in the Yenisei basin. Pages 5-7 *in* Materialy po issledovaniyu reky Yeniseia v rybopromyslovom otnoshenii. Krasnoyarsk. No. X.

Itra, A.R. and I.A. Veldre. 1988. On the distribution of some cancerogenic substances in water ecosystems. Page 33 *in* proc. of V Vsesoyuzn. konferentsiya po vodnoi toksikologii (V All-Union Conf. on Water Toxicology), Moscow.

Izyurova, A.I. 1955. The behavior of oil in water bodies. Gigiena i sanitaria 5: 15-18.

Kalashnikov, Yu. E. 1978. Fishes of the Vitim River Basin. Nauka Press, Novosibirsk. 190 pp.

Kaplina, G.S. 1974. Macrobenthos in the rocky beds of Lake Baikal littoral and its seasonal dynamics (data 1963-1968, Bolshiye Koty region). Pages 126-136 in Productivity of Baikal and anthropogenic changes of it's nature. Irkutsk.

Karantonis, F. E., F. N. Kirillov & F. B. Mukhomediyarov. 1956. Fishes of the middle Lena reaches. Trudy Instituta Biologii Yakutskogo Filiala AN USSR 2:3-144.

Karimov, B.K. 1989. On the Causes of Yearly Occurring Prespawning Mortality of Silver Carp and Bighead in Lake Tuzkan. Pages 244-245 in V11 Vsesoyuzn. konferentsia: Ekologicheskaya fiziologiya i biokhimiya ryb (VII All-Union Conf. on Fish Ecological Physiology and Biochemistry). Yaroslavl'.

Karolinskaya, Kh.M. 1952. Amitotic Division and its Role in the Cell Reproduction. Uspekhi sovremennoi biologii 33-2: 287-304.

Kasyanov V.P., V.P. Sirotkin and A.A. Khakimullin. 1979. On the reproduction of acipenserids in the Irtysh River basin. Pages 122-123 in proc. of II Vsesoyuzn. soveshchanie: Osetrovoe khozyaistvo vnutrennikh vodoemov SSSR (II All-Union Conf. on Sturgeon Farming in the USSR Inland Waters), Astrakhan.

Kazanskii, B.N. 1949. Functional features of the ovaries and pituitary of fish with portional spawning. Tr. Lab. osnov rybovodstva. 2: 64-120.

Kazanskii, B.N. 1979. Ecological-evolutionary principles of organization of sturgeon husbandry in the basins of the southers seas of the USSR. Pages 22-33 in Biological foundations for the development of sturgeon husbandry in the waters of the USSR. Nauka, Moscow.

Kessler, K.F. 1877. Fishes, Inhabiting and Encountered in the Aral-Caspian-Pontian Ichthyological Region, Trudy Aralo-Kasp. Ekspeditsii 4:312-313.

Khokhlova, L.V. 1953. The Chulym River: an essay on fisheries. Tr. Tomsk. Gos. Univ. ser. Biol. 125: 45-54.

Khokhlova, L.V. 1955. Sterlet (*Acipenser ruthenus natio Marsiglii* Brandt) in the Yenisei River. Voprosy Ikhtiologii (Issues in Ichtyology). 4:41-56.

Kiber. 1823. Notes on several elements of the natural history collected in Nizhne-Kolymsk and the surrounding areas in 1821. Pages 121-136 in G. Spassky (editor) Sibirskiy Vestnik (Siberian Investigator). Saint Petersburg. Book 10.

Kirillov, F. N. 1950. Fishes of the Tiksi Bay. Uchenye Zapiski Tomskogo Universiteta 15:155-162.

Kirillov, F. N. 1953. Fishes of the Indigirka River and their fishery. PhD Dissertation Thesis, Yakutsk. 13 pp.

Kirillov, F.N. Fish of the Indigirka River. 1955. Izvestiya VNII ozer. i rech. ryb. khozyayistva., 35, 141-167.

Kirillov, F.N. 1962. Ichthyofauna of the Vilyuy River basin. Pages 5-71 in Fauna of fish and invertebrates in the Vilyuy basin. Tr. Instituta biologii Yakutsk fil. Sib. Otd. AN SSSR, Yakutsk.

Kirillov, F. N. 1964. Species composition of fishes of the Aldan River. Pages 73-82 in Vertebrate Animals of Yakutiya, Yakutsk.

Kirillov, F.N. 1965a. Essential questions in fisheries related to the regulation of a river's flow.

Pages 68-76 *in* The nature of Yakutia and its protection. Trudy Instituta biologii Yakutskogo fililiala Sibirskogo Otdeleniya AN SSSR, Yakutsk.

Kirillov, F.N. 1965b. Biological foundations of fisheries in the water bodies of Eastern Yakutia. Pages 188-194 *in* The nature of Yakutia and its protection. Instituta biologii Yakutskogo fililiala Sibirskogo Otdeleniya AN SSSR, Yakutsk.

Kirillov, F. N. 1972. Fishes of Yakutia. Nauka, Moscow. 260 pp.

Kirillov, F.N. 1982. Small rivers - gene fund reserves of aquatic life. Page 9 *in* Problems of the Baikal area ecology. Part V *in* Genetic, Physiological and biochemical aspects of ecological montoring. Izd. BGNII, Irkutsk. 9 pp.

Kirillov, F.N. 1986. The effect of fisheries on the ichthyofauna of the Vilyuyskoye reservoir. Pages 39-40 *in* proceedings of the IX Symposium on Biologic problems of the North, Yakutsk. Yakutsky fililial Sibirskogo Otdeleniya AN SSSR, Yakutsk.

Kirillov, F.N. and V.S. Rybnikov. 1956. Fish and fisheries in the Vilyuy River. Pages 203-230 *in* Development of productivity in Western Yakutia in relation to the creation of diamond industry. Yakutsky fililial Sibirskogo Otdeleniya AN SSSR, Yakutsk.

Kirillov, S.D. 1991. Factors and results of antropogenic influences on the ichthyfauna of the upper Ob and the Novosibirskoye reservoir. Pages 132-136 *in* Ruboproduktivnost ozyor Zapadnoy Sibiri. SibRybNIIProyekt, Novosibirsk.

Knorre, A.G. 1959. Interrelationships Between Mitosis and Amitosis in the Courses of Individual Development and Evolution of Organisms. Tsitologiya 1: 494-509.

Kokarev A.V. 1983. Experiences of fish cultivation at the Tallin hatchery. Rybnoye Khozaistvo 2: 9-11.

Kokoza, A.A. 1970. Dynamics of Resistance against Phenol in the Early Ontogeny of Sturgeons. Pages 168-171 *in* Voprosy vodnoi toksikologii (Problems of Water Toxicology), Nauka, Moscow.

Kolzlyatkin A.L., C.K. Tyutenkov and L.P. Shendrik. 1973. Quantitative development and distribution of the zoobenthos in the Bukhtarminskoye reservoir (1967-1972). Pages 188-190 *in* Water-bodies of Siberia and perspectives for its exploitation for fisheries. Izdatelstvo Tomskoga Gosudarstvenogo Universiteta, Tomsk.

Komarovskii, F.Ya. 1987. Dynamics of Fish Poisoning by the Stable Pesticide. Pages 64-65 *in* I Vsesoyuzn. simpozium: Metody ikhtioroksikologicheskikh issledovanii (First All-Union Symp. on the Methods of Ichthyo-Toxicological Research). Leningrad.

Komarovskii, F.Ya., Kulik, V.A., Karasina, F.M. et al. 1988. Biological Indicators of Toxic Pollution in Water Ecosystems. Page 40 *in* V Vsesoyuzn. konferentsia po vodnoi roksikologii (V All-Union Conf. on Water Toxicology). Moscow.

Koshelev B. V. 1965. Regularities of generative cycle's changes of fishes. Pages 33-40 *in* Teoreticheskie osnovy rybovodstva. Nauka, Moscow.

Koshelev B. V. 1981. Study of fish reproduction (gametogenesis, rate of sexual maturity, reproductive cycles, spawning rhythms and ecology of spawning). Pages 5-16 *in* Issledovanija razmnozhenija I razvitija ryb. Nauka, Moscow.

Koshelev B. V. 1988a. The influence of anthropogenic factors on fish reproduction. Pages 51-59 in Metody bioindikatsii okruzhajushchei sredy v rajonakh AES. Nauka, Moscow.

Koshelev, B.V. 1988b. Pecularities of the Development and Functioning of Fish Reproductive Systems as an Indicator of Habitat Quality. Pages 41-42 in V Vsesoyuzn. konferentsia po vodnoi toksikologii (V All-Union Conf. on Water Toxicology). Nauka, Moscow.

Koshelev, B.V., G.I. Ruban, L.I. Sokolov, et al. 1989. Ecological and morphological characteristics of Siberian sturgeon from the middle and upper Lena basin. Pages 16-33 in Ekologiya, morfologiya i povedenie osetrovykh (Ecology, morphology, and behavior of sturgeon). Nauka, Moscow.

Kossov, M.F. 1933. Brief overview of the commercial fishery of YaASSR in 1927-30. Trudy Barabinskogo Otdeleniya Vsesoyuznogo Nauchno-Issledovatelskogo Instituta Ozernogo i Rechnogo Rybnogo Khozyaistva 4: 3-54.

Kovaleva, V.V. and Sergeeva, L.L. 1987. A Method for the Study of Toxicants Impact on Fish in the Food Chain Sediments-Benthos-Fish. Pages 56-57 in I Vsesoyuzn. simpozium po metodam ikhtiotoksikologicheskikh issledovanii (First All-Union Symp. on Methods of Ichthyo-Toxicological Research), Leningrad.

Kozhin N.I. 1949. Siberian sturgeon. *From Commercial Fish of the USSR* (atlas). Pischepromizdat. 787 pp.

Kozhov. M. M. 1950. Fresh waters of East Siberia. OGIZ Press, Irkutsk. 367 pp.

Kozhova, O.M., G.L. Okuneva, L.A. Izhboldina, G.S. Kaplina. 1974. The assessment of the distribution of the benthos in southern Baikal (Utilik-Murina regions) using Fisher's criteria. Pages 172-178 in Productivity of Baikal and anthropogenic changes of it's nature. BGNII, Irkutsk.

Krasnov, S.K. 1970. Impact of Elementary Phosphorus on Fish (Pathophysiological Study). Pages 125-128 in Voprosy vodnoi toksikologii (Problems of Water Toxicology), Nauka, Moscow.

Krayushkina, L.S. and S.N. Moiseenko. 1977. Functional Characteristics of Osmoregulation of Ecologically Distinct Groups of Sturgeons (Fam. *Acipenseridae*) in a Hypertonic Environment. Voprosy Ikhtiologii (Issues in Ichtyology). 17: 503-509.

Krivoshapkin M.F. 1865. The Yenisei region and its lifeforms. Saint Petersburg. Vol. 2. 188 pp.

Krokhalevskii V.P. 1997. Reproduction and stock of Siberian sturgeon in the Ob-Irtysh basin. Pages 116-117 in proceedings of Perviy kong. Ikhtiologov Rossii (First Congress of Russian Ichthyologists), Astrakhan. VNIRO, Moscow.

Kryazhev, A.I. and L.V. Chebasov. 1979. On the Problem of Fisheries Recapturing Coefficient Evaluation through the Marking of Juvenile Sturgeon. Pages 123-124 in II Vsesoyuzn. soveshchanie: Osetrovoe khozyaistvo vnutrennikh vodoemov SSSR (II All-Union Conf. on Sturgeon Farming in the USSR Inland Waters), Astrakhan.

Krylova, V. D. and L. I. Sokolov. 1981. Morphological Studies of Sturgeons and Their Hybrids (Recommendations on Methods). All-Union Research Institute for Sea Fisheries and Oceanography (VNIRO), Moscow. 49 pp.

Kuzmina S. S. 1979. The influence of some toxicants of sewage on gametes and early ontogenesis of wimba. PhD Thesis. Severtsev institute of animal morphology and ecology, Moscow. 24 pp.

Lapin, Yu. E., Yu. G. Yurovskii. 1959. On the interspecies patterns of maturation and dynamics of fecundity in fish. Zhurnal obschei biologii 20-6:47-56.

Lavrov S. 1909. Impressions and scienticfic results of a summer journey in the Yenisei tundra. Tr. Ob-va Estestvoispytateley pri Imperatorskom Kazanskom Universitete, Kazan'. vol. XLI. No 4 31 pp.

Lavrov S.D. and V.L. Issackenko. 1911. On the diet of fish in the lower Yenisei and the Yenisei Bay to Captain Varzugin Bay. Page 59 in Materials on the investigation of the Yenisei River with respect to fisheries. Kazan. 59 p.

Lavrova, T.V. and D.A. Chmilevskii. 1987. Impact of Increased Temperature on Ovogenesis in Tilapia (Oreochromis mossambicus Peters). Page 13 in Voprosy iskusstvennogo razvedeniya ryb. Vsesoyuznyi Nauchno-Issledovatelskyi Institut Ozernogo i Rechnogo Rybnogo Khozyaistva, Leningrad. p. 13.

Lemanova, N.A. 1955. An Analysis of Sterility in the Males of Ludoga (Coregonus lavaretus ludoga Pal.) and Ladoga (Coregonus infr. ladogensis Pravd.) Whitefish Hybrid, Dokl. Akad. Nauk SSSR 105: 160-162.

Lembik, Zh.L. 1976. On Benzopyrene Content in Some Water Objects. Pages 96-99 in Kantserogennye uglevodorody v promyshlennosti i okruzhayushchei cheloveka srede (Cancerogenic Carbohydrates in the Industry and Human Environment), Gor'kii.

Lepyoshkin, D.A. 1964. On the distribution of fish in the Viluy River basin. Pages 67-68 in Vertebrate animals of Yakutia. Isd. Institue Biologii SO AN SSSR, Yakutsk.

Lepyoshkin, D.A. 1970. Fish of the Olenek River and their commercial significance. Pages 113-114 in Voprosi Zoologii. Tomsk. gos. Univ., Tomsk.

Lesnikov, L.A. 1970. Peculiarities of the impact of pollution on populations of aquatic organisms. Pages 61-66 in Voprosy Vodnoi Toksikologii (Issues in hydro-toxicology). Nauka, Moscow.

Levin, N. 1899. The Fishery and Fishing Industry in the Lower Reaches of the Lena River. Izv. Vostochnosib. Otd. Imp. Rus. Geogr O-va 30-1: 91-129.

Logashov, M. V. 1940. Lake Melkoe and its utilization for fishery. Trudy Nauchno-Issledovatelskogo Instituta Polyarnogo Zemledeliya, Zhivotnovodstva i Promyslovogo Khozyaistva, Seriya Promyslovoe Khozyaistvo 11: 7-71.

Lukshene D. K. 1978. The influence of thermic regimes in the cooling reservoir of a Lithuanian power station on fish reproduction. Trudy Akademii nauk Litovskoi SSR. Ser. V. 4(84):81-91.

Lukshene, D.K., Yu. Virbitskas, and A. Astrauskas. 1979. The effect of water temperature in the Litovskaya Power Station reservoir on reproduction of bream and roach. Pages 128-129 in Ekologicheskiya fiziologiya i biokhimiya ryb (Ecological physiology and biochemistry of fish), Vol. 1. Astrakhan.

Lukyanchikov, F. V. 1966. Distribution og ichthyofauna in the Bratskoye reservoir in the period of its formation. Pages 51-52 in Sovesch. po biologitcheskoi produktivnosti vodoemov Sibiri (Conference on the biological productivity of Siberian water bodies) October 1966, Irkutsk - Abstracts. Izd. LIN CO AN SSSR, Irkutsk.

Lukyanchikov, F. V. 1967a. Fishes of the Khatanga River system. Trudy Krasnoyarskogo Otdeleniya SibNIIRKh 9:11-93.

Lukyanchikov, F. V. 1967b. Commercial and biological characteristics and condition of commercial fish in the Bratskoye reservoir in the first years of it's existence. Izv. Biol.-Geografich. NII of Irkutsk XX: 262-286.

Lukyanenko, V.I. 1967. Toksikologiya ryb (Fish Toxicology). Pishchevaya Promyshlennost, Moscow. 216 pp.

Lukyanenko, V.I. 1983. Obshchaya ikhtiotoksikologiya (General ichthyotoxicology). Legkaya i pishchevaya promyshlennost'. Moscow. 320 pp.

Lukyanenko, V.I. 1987a. Ekologicheskie aspekty ikhtiotoksikologii (Ecological Aspects of Fish Toxicology). Agropromizdat, Moscow. 239 pp.

Lukyanenko, V.I. 1987b. Current Problems of Fishery Toxicology as Related to the Problem of Water Quality Control. Pages 85-87 in I Vsesoyuzn. simpozium: Metody ikhtiotoksikologicheskikh issledovanii (First All-Union Symp. on the Methods of Ichthyo-Toxicological research), Leningrad.

Lukyanenko, V.I. 1990. The Impact of Multifactorial Anthropogenic Stress on the Living Conditions, Reproduction, Number, and Catches of Sturgeons. Pages 25-44 in Fiziologo-biokhimicheskii status volgo-kaspiiskikh osetrovykh v norme i pri rassloenii myshechnoi tkani (kumulyativnii politoksikoz) (Physiological and Biochemical State of Volgo-Caspian Sturgeons when Healthy and at Muscles Stratification (Cumulative Polytoxicosis)). Inst. Biol. Vnutrennikh Vod, Rybinsk.

Lukyanenko, V.I. 1992. How to Save the Caspian Sturgeons? Vestnik RAN 2: 55-74.

Lukyanenko, V.I., V.M. Raspopov and V.I. Dubinin. 1974. Seasonal dynamics of the gonadosomaticc index and its meaning for the foundation of rational harvest of the winter race of Russian sturgeon. Pages 90-91 in Otchetnaya Sessiya Tsentr. NII Osetr. Ryb. Khoz-va (Account Session of the Central Research Institute of Sturgeon Fishery), Astrakhan.

Maak, R. U. 1886. Vilyuy Region of the Yakutsk District, Part II. Saint Petersburg. 366 pp.

Maak, R. U. 1886. Vilyuy Region of the Yakutsk District, Part III. Saint Petersburg. 192 pp.

Makeeva, A.P. 1992. Embriologia Ryb (Fish Embryology). Moscow State University, Moscow. 216 pp.

Malyutin, V.S. 1965. Changes in the speed of embryogenesis in acipenserids as a function of temperature. Pages 40-48 in Zbornik Rabot po akklimitizattsii vodnikh organizmov. Pischevaia Promyschlenost.

Malyutin, V.S. 1980. Osobennosty ekologii lenskogo osetra i puti ego vosproizvodstva (Features of the Lena River sturgeon ecology and ways of its restocking). PhD. Thesis. Moscow. 159 pp.

Matkovskii, A.A. 1997. Changes in the principle factors acting on the Fish reserves of the Middle Ob. Page 122 in proceedings of Perviy kong. Ikhtiologov Rossii (First Congress of Russian Ichthyologists).

Mayr, E. 1969. Principles of systematic zoology. McGraw-Hill, New York. 428 pp. (Russian translation: Printsipy zoologicheskoi sistematiki, Mir, Moscow 1971).

Mazmanidy, N.D. 1970. On the pathomorphology of poisoning in fish by posphorous. Pages 128-136 in Voprosy vodnoy toksikologii (Issues is hydro-toxicology). Nauka, Moscow.

Meien V.A. 1940. On the reasons behind the variability in egg size in bony fish. Doklady AN SSSR (Academy of Sciences of the USSR) 28(7):654-656.

Meissner, V.I. 1933. Promyslovaya Ikhtiologia (Commercial Ichthyology). Snabtechizdat., Moscow-Leningrad. 192 pp.

Menshikov, M. I. 1936. On the biology of the Siberian sturgeon *Acipenser baeri* and the sterlet *Acipenser ruthenus* in the Irtysh River. Uchoniye Zapiski Permskovo G.U. (Scientific Reports of the Permsk State University) 2(1):41-65.

Menshikov, M. I. 1947. The geographical variation of the Siberian sturgeon, *Acipenser baeri* Brandt. Doklady AN SSSR., 55(4): 371-374.

Mikhailov, V.N. 1997. Ustia rek Rossii I sopredelnikh stran: proshloye , nastoyashee i buduschee (The mouths of the rivers of Russia and neighboring countries: past, present and future). GEOS, Moscow . 413 pp.

Mikhailova, L.V. 1991. Modern hydrochemical regimes and the effect of pollution an aquatic ecosustems and the fisheries of the Ob basin. Gidrobiol. Zhurnal 27(5):80-90.

Mikhailova, L.V., V.I. Uvarova and O.A. Barkhovich. 1988. Characteristics of the ionic content and mineralization of the waters in the Ob and some of its tributaries. Vodniye resursy 3:25-35.

Mikhalyov Yu. V. 1967. On the biology and regulation of the harvest of migratory sturgeon in the Yenisei River. Tr. Krasnoyarsk otd. Sib. NII rybn. khoz. IX:343-361.

Mikhalyov Yu. V. 1969. Reproduction of the Yenisei sturgeon. Pages 104-107 *in* Voprosy rybnogo khozaistva Vostochnoy Sibiri (Issues in fisheries management in Eastern Siberia) LIN SO AN SSSR, Irkutsk.

Mikhalyov Yu. V. 1991. Condition of the stock and perspectives for Yenisei sturgeon. Otchyot o NIR. Foundation of the Krasnoyarsk division of the East Siberian NII rybn. Khoz., Krasnoyarsk. 28 pp.

Mikhalyov Yu.V. et al. 1975. Biological foundations for the size of net mesh for catches of muksun, perch, omul and sturgeon. Pages 74-82 *in* Voprosy rybnogo khozaistva Vostochnoy Sibiri (Issues in fisheries management in Eastern Siberia) LIN SO AN SSSR, Irkutsk.

Mikheev, V.P. 1975. Russian and Siberian sturgeons as objects of aquaculture. Pages 103-104 *in* Osetrovoe khozyaistvo vodoemov SSSR (Sturgeon Management in USSR Waters), Astrakhan

Mikheev, V.P. 1979. Biological foundations of cage cultivation of fish in inland water bodies. Pages 168-184 *in* Biologicheskie resursi vnutrennikh vodoyomov SSSR (Biological resources of inland water bodies in the USSR). Nauka, Moscow.

Mikheev, V.P. 1982. Aquaculture of commercial fish. Lyogkaya I pischevaia promischlenost, Moscow. 216 pp.

Mikhin, V.S. 1941. Fish and fishereis of the Khatanga River and Khatanga Bay. Trudy Nauchno-Issledovatelskogo Instituta Polyarnogo Zemledeliya. Zhivotnovodstva i Promyslovogo Khozyaistva, Seriya Promyslovoe Khozyaistvo 16: 73-100

Miller, R. 1969. Freshwater fish of the Quaternary in North America. Pages 174-192 *in* The Quaternary Period in the USA. MIR, Moscow.

Mina, M.V. 1986. Mikroevolyutsiya ryb: Evolyutsionnye aspekty feneticheskogo raznoobraziya (Microevolution of Fish: Evolutionary Aspects of Phenetic Diversity), Nauka, Moscow. 207 pp.

Morozov, V.A. 1951. On the divergence in growth of young fish and reasons for these divergences. Zool. Zhurnal 30(5):457-465.

Moor, G.V. and S. Ramamurthi. 1951. Tyazhyoliye metally v prirodnikh vodakh (Heavy Metals in Natural water bodies). Mir, Moscow. 286 pp.

Natochin, Yu.V, V.I. Lukyanenko, E.I. Shakhmatova, et al. 1995. Twenty Years of Monitoring (1970-1990s) of the Physico-chemical Characteristics in the Blood Serum of the Russian Sturgeon *Acipenser gueldenstaedti*. Voprosy Ikhtiologii (Issues in Ichtyology) 35(2):253-257.

Naumov, N.P. 1964. On Methodological Problems of Biology. Nauch. Dokl. Vyssh. Shkoly. Filosofskie Nauki 1:26-34.

Nesov, L.A. and M.N. Kaznyshkin. 1983. New sturgeons from Cretaceous and Paleogene of USSR. Pages 68-76 *in* V.V.Menner (ed.) Sovremenyie problemy paleoikhtiologii (Current Issues in Paleoichthyology). Nauka, Moscow.

Nikolskaya, N.G., and L.A. Sytina. 1974. Zone of temperature adaptations during the Lena River sturgeon eggs development. Pages 108-109 *in* T.V. Astakhova, editor. Tezisy otchetnoi sessii Tsentralnogo Nauchno-Issledovatelskogo Instituta Osetrovogo Rybnogo Khozyaistva (VNIRO).

Nikolskii, A. M. 1896. Siberian sturgeon *(Acipenserstenorrhynchus sp. nov.)*. Ezhegodnik Zoologicheskogo Museya Akademii Nauk 1: 400-405.

Nikolskii, A.M. 1902. Fish and Reptiles. Izd. Brokgauz-Efron, Petrograd. 872 pp.

Nikolskii, G. V. 1939. Materials on taxonomy of the Siberian sturgeon, *Acipenser baerii* Brandt. Sbornik Trudov Gosudarstvennogo Zoologicheskogo Muzeya pri MGU 5:136-148.

Nikolskii, G.V. 1971. Chastnaya Ikhtiologiya (Special Ichthyology). Visshaya Shkola, Moscow. 471 pp.

Nikolskii, G.V. 1974. Teoriya dinamiki stada ryb kak biologicheskaya osnova ratsionalnoi expluatatsii i vosproizvodstva rybnykh resursov (Theory of fish stock dynamics as a biological base of rational exploitation and restocking of fish resources). Pishchevaya promyshlennost, Moscow. 447 pp.

Novikov, A. S. 1966. Ryby reki Kolymy (Fishes of the Kolyma River). Nauka, Moscow. 134 pp.

Novikov, G.G. 1991. Osobennosty rosta i energetiki razvitiya kostistykh ryb v rannem ontogeneze (Features of growth and developmental energetics of teleostean fishes during early onthogenesis). Abstract of doctors dissertation. Moscow. 44 pp.

Novosadova, T.G. 1985. Control of toxic runoff from oil-refineries. Pages 48-50 *in* Tekhnologiya fiziko-khimicheskoi ochistky promyshlenykh stochnykh vod analiticheskiy control I regulirovaniye protsessov ochistky (Technology of physico-chemical purification of industrial runoff). Trudy NII VodGeo, Moscow.

Orlova, G.A. and V.M. Shirokov. 1976. Hydrological regime of the Novesibirskoye reservoir. Pages 5-15 *in* Biologischeskiy Rezhiim i rybkhozaistvenovo ispolzovaniye Novosibirskogo vodokhranilischa (Biological regime and the use of the Novosibirskoye reservoir for fisheries). Zapadno-Sibirskoye knizhnoye izdatelstvo, Novosibirsk.

Ostroumov, N. A. 1937. Fishes and fishery of the Pyasina River. Trudy Polyarnoi Komissii AN USSR 30:1-115.

References

Pankratova, V. Ya. 1970. Larvae of mosquitos in the subfamily *Orthocladiinae*. From Fauna SSSR. Nauka, Leningrad. 343 pp.

Pavlov, A.V. and V.A. Raspopov. 1972. Analysis of the spawning grounds of sturgeon and beluga in the Volga River in 1970. Pages 86-88 *in* Materialy obyedineniye nauchnoi sessii Tzentralnogo nautshno-issledovatelskogo instituta osetrovogo rybnogo khosjaistva. Astrakhan.

Pavlov, D.S., K.A. Savvaitova, L.I. Sokolov and S.S. Alekseev, 1994, Redkie i ischezayushchie zhivotnye. Ryby (Rare and Endangered Animals. Fish), Vysshaya Shkola, Moscow. 334 pp.

Pertseva, T.A. 1939a. Materials on the development of the Caspian puzanok *Caspioalosa caspia*. Trudy All-Union Research Institute for Sea Fisheries and Oceanography (VNIRO). 8: 27-61.

Pertseva, T.A. 1939b. Spawn, larvae and fry of fish in the Motov Bay. Trudy All-Union Research Institute for Sea Fisheries and Oceanography (VNIRO). 4: 417-467.

Petkevich, A.N. 1952. Biology and reproduction of sturgeon in the Middle and Upper Ob related to hydroconstruction. Trudy. Tomsk. Univ. 119:39-64.

Petkevich, A.N. 1953. On the Morphology of Siberian Sturgeon, Trudy Barabinskogo Otdeleniya Vsesoyuznogo Nauchno-Issledovatelskogo Instituta Ozernogo i Rechnogo Rybnogo Khozyaistva 6(2):3-16.

Petkevich, A. N., V. N. Bashmakov & A. Ya. Bashmakova. 1950. Sturgeons of the middle and upper reaches of the Ob River. Trudy Barabinskogo Otdeleniya Vsesoyuznogo Nauchno-Issledovatelskogo Instituta Ozernogo i Rechnogo Rybnogo Khozyaistva 4:3-54.

Petkevich, A. N. and Iogansen B.G. 1958. Perspectives for fisheries in the upper Ob related to hydroconstruction. Izvestiya Vsesoyuznogo Nauchno-Issledovatelskogo Instituta Ozernogo i Rechnogo Rybnogo Khozyaistva 44:5-28.

Pezhemskii, P.P. 1853. Fish productivity of Lake Baikal. Vestnik Imperatorskogo Russkogo Geograficheskogo Obshchestva 8: 9-34.

Pirozhnikov, P.L. 1955. Materials on the Biology of Commercial Fishes of the Lena River, Izvestiya Gosudarstvennogo Nauchno-Issledovatelskogo Instituta Ozernogo i Rechnogo Rybnogo Khozyaistva. 35:61-128.

Pirozhnikov, P.L. 1959. Fauna complexes and ecological classification of fish in the Lower Lena. Pages 91-100 *in* Biologichedskiye osnovi rybnogo khozaistva (Biological foundations of fish management). Izdatelstvo Tomskogo Gosudarstvenogo Universiteta, Tomsk.

Plokhinskii, N.A. 1970. Biometriya (Biometry). Mosk. Gos. Univ., Moscow. 368 pp.

Podlesnyi, A.V. 1954. Spawning Migrations of Yenisei Anadromous Fishes in Relation to the History of Yenisei. Zoologitcheskii Zhurnal. 33(1):120-126.

Podlesnyi, A. V. 1955. Sturgeon (*Acipenserbaerii stenorrhynchus* A. Nikolsky) of the Yenisey River. Voprosy Ikhtiologii (Issues in Ichtyology) 4: 21-40.

Podlesnyi, A. V. 1958. Fishes of the Yenisey River, conditions of their life and utilization of them. Trudy Vsesoyuznogo Nauchno-Issledovatelskogo Instituta Ozernogo i Rechnogo Rybnogo Khozyaistva 44: 97-178.

Podlesnyi, A. V. 1963. State of stock of sturgeons in the Yenisey River and ways to increase them. Pages 200-205 *in* E. N. Pavlovskii (ed.)

Sturgeon Fishery in Water Bodies of the USSR. Izdatelstvo Akademii Nauk USSR, Moscow.

Podlesnyi, A.V. 1968. Principal differences of migratory bony fish from non-migratory fish. Voprosy Ikhtiologii (Issues in Ichtyology) 8(2):211-215.

Pomazovskaya, I.V., E.V. Flink, and L.V. Dubrovina. 1988. The Role of Abiotic Factors in the Development of Intoxication in Aquatic Animals. Page 66 in V Vsesoyuzn. soveshchanie po vodnoi toksikologii (V All-Union Conf. on Water Toxicology), Moscow.

Popova G. V. 1978. The abnormalities in fish oogenesis under the influence of some pesticides. Pages 137-142 in Izmenenie prirodnoi sredy v svyazi c deyatelnostju tsheloveka (Changes in the natural environment due to the activity of man). Moscow.

Popova, G.V. 1987. Methodological Aspects of the Studies on Bioaccumulation of Pesticides in Fish. Pages 111-113 in I Vsesoyuzn. simpozium: Metody ikhtiotoksikologicheskikh issledovanii (First All-Union Symp. on the Methods of Ichthyo-Toxicological Research), Leningrad.

Ramazanov, P.N. 1966. Pollution in fished water bodies of the Krasnoyarsk district. Pages 171-172 in Proceedings on the biological productivity of Siberian water bodies (October 1966, Irkutsk). Short Reports LIN SO AN SSST, Irkutsk.

Raspopov, V.M. 1987. Fecundity of the Great Sturgeon Huso huso. Voprosy Ikhtiologii (Issues in Ichtyology). 27(2):254-263.

Rasputin, I.O. 1903. On the fisheries of the Lower Kolyma. Vestnik rybopromishlenesti. 3: 99-105.

Revnivikh, A.I. 1937. On the diet of acipenserids and salmonids in the Irtysh basin. Trudy Permskogo Biol. Inst. VII(3-4):261-282.

Reznichenko, P.N., I.N. Ziborova and V.S. Malyutin. 1979. Effect of constant incubation temperatures on the development of eggs of the Siberian sturgeon in the Lena River. Ekologisheskaya fiziologiya I biokhimiya ryb. II: 162-164.

Rikhvanov, L.P. 1996. Radioecological situation on the territory of the Ob River basin. Pages 23-27 in Radioactivity and radioactive elements in inhabited environments. Izd. Tomsk. Politekh. Univ., Tomsk.

Rogovskaya, Ts.I. 1967. Biochemical methods for the purification of industrial run off. Stroyizdat, Moscow. 14 pp

Romanenko, T.I. and N.F. Tryamkina. 1978. Hydrobiological regimes and productive abilities of the Ust'-Ilimskoye reservoir. Pages 253-256 in Productivity of water bodies of various cliamtic zones in the RSFSR and perspectives for commercial fisheries. Izd. SibRybNIIproyekt, Krasnoyarsk.

Romanov, A.A. 1990. Anomalies in Morphogenesis of the Gonads, Reproductive Cells, and Livers of Caspian Sturgeon during their Marine Life. Pages 83-85 in Simpozium po ekologicheskim i morfologicheskim osnovarn adaptatsii gidrobiontov (Symp. on Ecological and Morphological Essentials of Adaptations in Aquatic Organisms). Leningrad.

Romanov, A.A. and N.N. Sheveleva. 1992. Anomalies of Gonadogenesis in the Caspian Sturgeon (*Acipenseridae*). Voprosy Ikhtiologii (Issues in Ichtyology). 32(5):176-180.

Romanov, A.A. and N.N. Sheveleva. 1993. Anomalies of Morphogenesis in Caspian Sturgeons. Rybn. Khoz. 4:27-28.

Romanov, A.A. and Yu.V. Altufiev. 1990. Neoplasms in the Gonads and Livers of the

Sturgeon (*Acipenseridae*) from the Caspian Sea. Voprosy Ikhtiologii (Issues in Ichtyology). 30(6):1040-1044.

Romanov, A.A. and Yu.V. Altufiev. 1992. The Extraregional Histogenesis of Gametes in Sturgeon from the Caspian Sea. Voprosy Ikhtiologii (Issues in Ichtyology) 32(5):145-154.

Romanov, A.A., N.N. Sheveleva, N.N. and Yu.V. Altufyev. 1990. Anomalies of Gonado- and Gametogenesis in Sturgeon from the Caspian Sea. Pages 92-100 *in* Fiziologo-biokhimicheskii status volgokaspiiskikh osetrovykh v norme i pri rassloenii myshechnoi tkani (kumulyativnyi politoksikoz) (Physiological and Biochemical State of Volgo-Caspian Sturgeons when Healthy and at Muscles Stratification (Cumulative Polytoxicosis)). Inst. Biol. Vnutrennikh Vod, Rybinsk.

Romanov, A.A., S.E. Zubova, L.V. Piskunova, et al. 1984. Characteristics of the Early Period in the Development of Great Sturgeon Gonads during its Marine Life. Pages 300-302 *in* Osetrovoe khozyaistvo vodoemov SSSR (Sturgeon Management in the USSR Water Bodies), Astrakhan.

Romanova, G. P. 1948. Feeding of fishes in down streams of the Yenisei River. Trudy Sibirskogo otdeleniya Vsesoyuznogo nauchno-issledovatelskogo instituta ozernogo i rechnogo rybnogo khozyaistva. Krasnoyarsk 2:151-203.

Roskin G. I., L.B. Levinson. 1957. Mikroskopicheskaya tekhnika (Microscopic technique). Moscow, Sovetskaja nauka. 468 pp.

Roskin, G.I. and M.E. Struve. 1948. Cytological Differences between Mitosis and Amitosis. Dokl. Akad. Nauk SSSR. 59(1):143-145.

Ruban, G. I. 1986. Morphological variation in the Siberian sturgeon, *Acipenser baeri,* of the Lena River as effect of warm-water cultivation. Voprosy Ikhtiologii (Issues in Ichtyology) 26(3): 470-475.

Ruban, G. I. 1989a. Clinal variation of morphological characters in the Siberian sturgeon, *Acipenser baeri,* of the Lena River basin. Voprosy Ikhtiologii (Issues in Ichtyology) 29(5): 718-726 (English translation: J. Ichthyol. 29: 48-55).

Ruban, G. I. 1989b. Morphological variation in the Siberian sturgeon of the Lena basin. Pages 5-16 *in* Morfologiya, ekologiya i povedenie osetrovikh (Morphology, ecology and behavior in acipenserids). Nauka, Moscow.

Ruban, G.I. 1995a. Species Structure and Population Status of the Siberian Sturgeon Acipenser baerii. Population ecology: structure and dynamics. Proceedings of the Obsche-Rossiyskogo Sovescheniye (15-18 November 1994, Pushchino). IPEE RAN, Moscow.

Ruban, G. I. 1995b. Some factors influencing the shape of the growth curve in northern Lena sturgeon. Ekosistemy Severa: struktura, adaptatsia, ustoichivost. Pages 99-103 *in* Proceedings of the Obsche-Rossiyskogo Sovescheniye (26-28 October 1993, Petrosavodsk). Isd. Nauch. Coveta po Problem Ekologicheskich Sistem (1995b).

Ruban, G. I., and N.V. Akimova. 1991. The ecological features of the Siberian sturgeon Acipenser baerii Brandt from the Indigirka River. Voprosy Ikhtiologii (Issues in Ichtyology) 31(4): 596-605 (English translation: J. Ichthyol. 31:118-129).

Ruban, G. I., and N.V. Akimova. 1993. Ecological Characteristics of Siberian Sturgeon, Acipenserbaeri, from the Kolyma River.

Voprosy Ikhtiologii (Issues in Ichtyology) 33(1): 84-92.

Ruban, G. I. and L.A. Konoplya. 1994. Diet of the Siberian Sturgeon, *Acipenser baeri,* in the Indigirka and Kolyma rivers. Voprosy Ikhtiologii (Issues in Ichtyology) 34(1): 130-132,

Ruban, G. I. and A. I. Panaiotidi. 1994. Comparative morphological analysis of subspecies of the Siberian sturgeon, *Acipenser baerii stenorrhynchus* A. Nikolsky and *Acipenserbaerii chatys* Drjagin (Acipenseriformes, Acipenseridae), in the Yenisey and Lena rivers. Voprosy Ikhtiologii (Issues in Ichtyology) 34(4): 469-478 (English translation: J. Ichthyol. 34: 58-71).

Ruban, G. I. and L. I. Sokolov. 1986. Morphological variation of Siberian sturgeon (*Acipenser baeri*, Brandt) of the Lena in connection with its rearing in warm water. Voprosy Ikhtiologii (Issues in Ichtyology) 26(3): 470-475.

Rustamova, Sh.A. and Kasimov, R.Yu. 1968. The Impact of Oil from the Field "Neftyanye Kamni" on Various Concentrations of Fish and their Prey. Pages 33-35 *in* Razrabotka biologicheskikh osnov i biotekhniki razvitiya osetrovogo khozyaistva v vodoemakh SSSR (po materialam 1967 g.) (Development of Biological Bases and Biotechnology for Organization of Sturgeon Management in the USSR Water Bodies), Astrakhan.

Ruzkiy, M.D. 1916. On fish from the upper Yenisei. Izv. Imperator. Tomsk. Univ. 65:1-18.

Ryby Kazakhstana (Fish of Kazakhstan) vol. 1. 1986. Nauka, Alma-Ata. 272 pp.

Rybnikov, V.S. 1961. Condition and ways of development for fisheries in the Yana River. Pages 125-131 *in* Voprosy geographii Yakutii. Izd. Yakutsk. fil. Sib. Otd. AN SSSR, Yakutsk.

Sakun, O.F. 1954. Histological Analysis of the Male Gonads in Vimba (*Vimba vimba* L.) Following Spawning. Dokl. Akad. Nauk SSSR 98(4): 657-659.

Sakun O. F. 1964. The amitosis of fish oocytes and conditions of its appearance. Pages 295-297 *in* Problemy sovremennoi embriologii. Moscow.

Sakun, O.F. and V.G. Svirskii. 1992. Degeneration of Oocytes during the Previtellogenesis and Vitellogenesis in the Sexual Cycle of Pacific Sardine *Sardinops sagax melanosticta.* Voprosy Ikhtiologii (Issues in Ichtyology) 32(3):52-58.

Saldau, M.P. 1948. Diet of the fish of the Ob-Irtysh basin. Izvestiya Vsesouznogo Instituta Ozernogo i Rechnogo Rybnogo Khozyaistva 28: 175-226.

Savvaitova, K.A., M.I. Pichugin and V.A. Maksimov, S.V. Maksimov and S.D. Pavlov. 1994. Chagnes in the composition of the ichthyofauna of the water bodies of the Norilo-Pyasinskii water system under conditions of intensive anthropogenic influence. Voprosy Ikhtiologii (Issues in Ichtyology) 34(4):566-569.

Savvaitova, K.A., Yu.V. Chebotareva, M.Yu. Pichugin, and S.V. Maksimov. 1995. Abnormalities in Fish Morphology as Indicators of Environment Quality. Voprosy Ikhtiologii (Issues in Ichtyology) 35(2):182-188.

Sedelnikov, A. K. 1910. Lake Zaisan. Zapiski Zapadno-Sibirskogo Otdela Imperatorskogo Russkogo Geograficheskogo Obshchestva 35:1-253.

Selyukov, A.G. 1997. Condition of the reproductive system and liver of fish in conditions of intense pollution in the Ob-Irtysh basin. Pages 116-117 *in* Proceedings of the First Ichthyological

Congress of Russia. Astrakhan.

Selyukov, A.G. and A.M. Stepanov. 1988. Oogenesis and reproductive potential of the Ob muksun in modern conditions. Pages 24-26 in Puty povisheniya produktivnosti i ratsionalnoye ispolzovanie rybnikh resursov vnytrenikh vodoemov. SibRybNIIProjekt, Tyumen.

Semyonov, K.I. 1963. Biological variability in quality of sturgeon eggs and it's influence on the development of the larvae in artificial rearing. Voprosy Ikhtiologii (Issues in Ichtyology) 3(26):99-113.

Semyonova, L.A., N.S. Knyazeva, V.B. Stepanova, A.I. Kovalenko and T.V. Zakharova. 1997. Contemporary condition of the environment of fish in the Lower Ob. Pages 171 in in Proceedings of the First Ichthyological Congress of Russia. Astrakhan.

Serebryakova, E.V. 1964. A Study of the Gonads of Spawning Sturgeon from the Volgograd Reservoir. Tr. Vses. Inst. Morsk. Rybn. Khoz. Okeanograf. (Proc. All-Union Research Institute for Sea Fisheries and Oceanography (VNIRO)), Moscow. 56(3):117-130.

Seryshev, V.A. and A.A. Ozherelyev. 1978. Impact of flooded soils on gaseos content of water in the Ust-Ilimskoye reservoir. Pages 260-262 in Produktivnost' vodoyomov raznikh klimaticheskikh zon RSFSR i perspektivi. SibRybProyekt, Krasnoyarsk.

Seroschevskii, V.L. 1896. Yakuty: Opyt Etnograficheskogo issledovaniya (Yakutians, Ethnographic study) Saint Petersburg. 720 pp.

Setsko, R.I. 1976. Acipenserids and nelma in the Novosibirskoye reservoir. Pages 126-133 in Biologicheskiy rezhim I rybokhozaistvonnoye ispolzovaniye Novosibirskogo vodokhraniliscsha. Zap. Sib. Knizh. Izd. Novosibirsk.

Shagaeva, V.G., N.G. Nikolskaya, K.P. Markov, et al. 1989. Peculiarities of embryonic and larval development of sturgeon under deteriorating ecological conditions in the Volga River. Pages 336-337 in The Sturgeon Fishery in USSR Waters. Kratkie tezisy nauch. dokl. k predstoyashchemy Vsesoyuz soveshch., Part 1. Astrakhan.

Shagaeva, V.G., N.V. Akimova, K.P. Markov, et al. 1991. Pathological changes in the early ontogeny of acipenserid fishes in the Volga River when exposed to human influences. Pages 160-161 in Fifth All-Union Conference on Early Ontogeny of Fish. Abs. of Reports, Astrakhan.

Shagaeva, V.G., Nikol'skaya, M.P., Akimova, N.V. et al. 1993. A Study of the Early Ontogenesis in Volga Sturgeons (Acipenseridae) in Connection with an Anthropogenic Impact. Voprosy Ikhtiologii (Issues in Ichtyology) 33(2):230-240.

Shakhmaev, N.K. 1973. Freshwater mollusks as a bioindicator of magnesium, cobalt, copper and iron concentrations. Pages 217-218 in Vodoyemi Sibirri I perspektivi ikh rybokhozaistvenogo ispolzovaniya. Tomsk Univ, Tomsk.

Shakhmatova, R.A. 1983. Adaptation of Some Aquatic Organisms to the Waste Waters of Chemical Industry. Pages 117-121 in Reaktsiya gidrobiontov na zagryaznenie (Response of Aquatic Organisms on Pollution), Nauka, Moscow.

Shatunovskii, M.I. 1980. Ekologicheskiye zakonomernosti obmena veschestv u morskikh ryb (Ecological appropriatenesses in marine fish metabolism). Nauka, Moscow. 283 pp.

Shatunovskii, M.I. 1986. On some metabolic adaptations of fishes at the north borders of their range. Pages 143 - 144 *in* V. E.Sokolov, Yu.I.Chernov and B.Ya.Vilenkin, editors. Tezisy doklada na Vsesoyuznom soveshchanii Organizmy, populatsii i soobshchestva v ekstremalnykh usloviyakh, Moscow.

Shatunovskii, M.I., N.V. Akimova, G.I. Ruban. 1996. Responses of the reproductive system in fish on anthropogenic influences. Voprosy Ikhtiologii (Issues in Ichtyology) 36(2):229-238.

Shelukhin, G.K., G.F. Metallov, P.P. Geraskin et al.1989. Indexes of the Metabolism in Russian Sturgeon with Different Degree of Muscles Pathology during their Marine Life. Pages 345-347 *in* Osetrovoe khozyaisrvo vodoemov SSSR (Sturgeon Farms of USSR Water Bodies), Astrakhan.

Sheveleva, N.N. 1990. On the Disturbance of Gametogenesis in Caspian Sturgeon under Present-day Conditions. Pages 107-108 *in* Ekologicheskie i motofunktsionalnye osnovy adaptatsii gidrobiontov (Ecological and Morphofunctional Bases of Adaptations in Aquatic Organisms), Leningrad.

Shmalgauzen, I.I. 1961. The Integrity of Biological Systems and their Self-Regulation, Bull. Mosk. Obshch. Ispyt. Prir., Ord. Biol., 66(2):104-134.

Shmalgauzen, I.I. 1968. Faktory evolyutsii. Teoriya stabiliziruyushchego otbora (Factors of Evolution: The Theory of Stabilizing Selection), Nauka, Moscow. 451 pp.

Shmalgauzen, I.I. 1982. Organizm kak tseloe v individual'nom i istoricheskom razvitii (Organism as a Whole during Ontogeny and Evolution), Nauka, Moscow. 228 pp.

Shmidt, P.Yu. 1936. Migratsii ryb (Fish Migrations), Biomedgiz, Moscow. 327 pp.

Shostakovich, V.B. 1909. Materials on the climatology of the Russian Orient, Part 1: Covering and Freezing the waters in the Russian Orient. Izv. Vost. Sib. otd. Russ. Geog. Obshch vol. 37.

Shtylko, B.A. 1934. The Neogene fauna of freshwater fishes in West Siberia. Trudy Vsesoyuznogo geologorazvedochnogo obyedineniya 359: 7-23.

Shubnikov, D.A. 1976. Types of Migration Cycles in Anadromous and Semianadromous Fish. Voprosy Ikhtiologii (Issues in Ichtyology) 16(4):587-591.

Shulga, E.L. 1967. Some data on the zooplankton of the Bratskoye reservoir. Izv. Biol. Geog. NII, Irkust Gos. Univ, Irkutsk. 20: 110-117.

Shumaelov, A.V. 1986. Reproduction and protection of fish stocks in the lower section of the Angara River in the zone of future Bogodanskoye reservoir. Pages 72-73 *in* Ikhtiologia, gidrobiologia, gidrokhimiya, entomologiya i parasitologiya. IX Vsesoyus. Symp. Biolog. Problem Severa. Institut Biologii YaF SO AN SSSR Irkutsk.

Shvarts,S.S.1980.Ekologicheskiezakonomernosti evolyutsii (Ecological Regularities of Evolution), Nauka, Moscow. 280 pp.

Shylov, V.I. 1964. Maturation and periodicity of spawning of sterlet in the Volgogradskoye reservoir. Trudy All-Union Research Institute for Sea Fisheries and Oceanography (VNIRO). 56(3):79-104.

Skalon, V.N. 1931. Fisheries in the region of the Taz River. Sovetskiy Sever. 9:11-16.

Smirnov, M.P. 1975. Geographic distribution, regional characteristics of organic substances and mineralization of river waters in the middle taiga of the USSR. Gidrokhimicheskiye

Materialy 42:20-41.

Sokolov, B.E. and M.I. Shatunovskii (ed.) Ryby Mongolskoi Narodnoi Respubliki (Fish of the Peoples Republic of Mongolia). Nauka, Moscow. 277 pp.

Sokolov, L.I. 1965a. Growth of Siberian sturgeon *Acipenser baerii* Brandt of the Lena River. Vestnik MGU. 1:3-12.

Sokolov, L.I. 1965b. Maturation and fecundity of the Siberian sturgeon *Acipenser baeri* Brandt from the Lena River. Voprosy Ikhtiologii (Issues in Ichtyology) 5(1): 70-81.

Sokolov, L.I. 1966a. Siberian Sturgeon *Acipenser baerii* Brandt of the Lena River. PhD Dissertation. Moscow State Univ., Moscow. 170 pp.

Sokolov, L.I. 1966b. Diet of the Siberian sturgeon *Acipenser baerii* Brandt of the Lena River. Voprosy Ikhtiologii (Issues in Ichtyology) 6(3):550-560.

Sokolov, L.I. 1981. Ecology of the Lena sturgeon and perspectives for its fishery. Pages 211-237 *in* Ekologo-faunistichiskiye issledovaniya - Biologicheskiye Resursy Territory v zone BAM. MGU, Moscow.

Sokolov, LI., and N.V. Akimova. 1976. A technique for determining the age of Siberian sturgeon of the Lena River. Voprosy Ikhtiologii (Issues in Ichtyology) 16 (5): 853-858.

Sokolov, L. I., and S. M. Kashin. 1965. Comparative analysis of some morphobiological indicators in populations of the Siberian sturgeon (Acipenser baeri, Brand) from various waters. Vestn. Mosk. gos. un-ta, 3:13-18.

Sokolov, L. I., B. V. Koshelev, O.V. Khalatyan, G. I. Ruban, N. V. Akimova, E. L. Sokolova. 1984. Ecological and morphological characteristics of the Siberian sturgeon (*Acipenser baerii* Brandt) from the Aldan River. Pages 333-334 in Proceedings of the Vsesoyuz. Covesch. Astrakhan.

Sokolov, L. I., B. V. Koshelev, O.V. Khalatyan, at al. 1986. Ecological and morphological characteristics of the Siberian sturgeon (*Acipenser baerii* Brandt) in the Aldan. Voprosy Ikhtiologii (Issues in Ichtyology) 26(5): 741-749.

Sokolov, L.I., and Malyutin, V.S. 1977. The Structure of the Population in the Siberian Sturgeon from River Lena at Spawning Sites. Voprosy Ikhtiologii (Issues in Ichtyology) 17(2):327-246.

Sokolov, L.I., and A.S. Novikov. 1965. Data on the biology of the Siberian sturgeon (*Acipenser baeri* Brandt) in Yakut waters. Nauch. dokl. vyssh. shk. Ser. biol. nauk. 4:36-38.

Sokolov, L.I., and G.I. Ruban. 1979. Different properties among female Siberian sturgeon (*Acipenser baeri hatys* Drjagin) in the Lena River and some indicators of their reproductive capacity. Byul. MOIP. Otd. biol., 84(6): 67-73.

Sokolova, Ye.M. 1951. Thermal regimes of the rivers of the USSR. Trudy Gosudarstvennogo Gidrologicheskogo Instituta 31:1-116.

Solomonovskaya, V.P. 1952. Diet of several fish in the upper and middle Ob. Trudy Tomsk Gos. Univ. 119:65-74.

Solovov, V.P. 1997. Contemporary condition of Siberian sturgeon populations *Acipenser baeri* in the Upper Ob. Voprosy Ikhtiologii (Issues in Ichtyology) 37(1):47-53.

Statova, M.P. 1985. Comparative ecological-morphological research on several carp from waters of Moldavia. Pages 99-111 *in* Osobennosti reproduktivnykh tsiklov u ryb v vodoyemakh raznykh shirot (Features of the reproductive cycle of fish from various latitudes). Nauka Press, Moscow.

Stroganov, N.S. 1970. Water pollution and issues by hydrotoxicology. Pages 11-23 in Voprosy Vodnoy Toksikologii. Nauka, Moscow.

Stroganov, N.S. 1971a. A Method for Assessment of Water Toxicity. Pages 14-60 in Metodiki biologicheskikh issledovanii po vodnoi toksikologii (Methods of Biological Studies Related to Water Toxicology), Nauka, Moscow.

Stroganov, N.S. 1971b. A Biological Criterion of Toxicity in Water Toxicology. Pages 14-28 in Kriterii toksichnosti i printsipy metodik po vodnoi toksikologii (Criteria of Toxicity and Bases of Techniques in Water Toxicology). Mosk. Gos. Univ., Moscow.

Stroganov, N.S. and Pozhitkov, A.T. 1941. The Impact of Industrial Waste Waters on Aquatic Organisms (New Ways to Solve the Problem). Uch. Zap. Mosk Gos. Univ 60:88.

Stroganov, N.S. and Telitchenko, M.M. 1958. The Chronic Impact of Low Doses of Uranium-238 on the Bleak Gonads. Bull. Mosk. Obshch. Ispyt. Prir. Ord. Biol. 63(4):154-159.

Stroganova, N.S. 1952. Amitosis at the Spermatogenesis in Mammals. Dokl. Akad. Nauk USSR, 85(4):897-900.

Stygar, I.E., Gaponov, V.S. and Drizo, E.A. 1979. The Impact of Artesian Water and Some Components on the Motility and Fertilization Ability of the Sperm of Sturgeon. Pages 247-248 in Osetrovoe khozyaisrvo vnutrennikh vodoemov SSSR (Sturgeon Management the USSR Inland Waters). TsNIORKh Astrakhan.

Stygar, I.E., Gaponov, V.S. and Drizo, E.A. 1981. The Impact of Artesian Water and Some Components on the Early Stages of Sturgeon Development. Pages 224-225 in Ratsional'nye osnovy vedeniya osetrovogo khozyaisrva (Rational Bases of Management of Sturgeons Farms). TsNIORKh Volgograd and Astrakhan.

Suvorov, E.K. 1948. Osnovy ikhtiologii (Fundamentals of Ichthyology). Sov. Nauka, Moscow. p. 580.

Sytnik, Yu.M. 1992. Accumulation of Strontium-90 and Cesium137 in the Components the Kiliiskaya Delta Ecosystem of the Danube River. Cand. Sc. (Biol.) Dissertation, Kiev. 19 pp.

Tatarko, K.I. 1977. Anomalies in carp and the role of temperature in their development. Pages 157-196 in Biologicheskiy rezhim vodoyemov-okhladiteley TETs i vliyaniye temperatury na gidrobiontov (The biological regime of cooling waters for thermal power plants and the effect of temperature on aquatic life). Nauka Press, Moscow.

Terian, T.G. 1942. The Impact of High Temperature on the Sex Differentiation and Redifferentiation of Fish. Bull. Exp. Biol. Med. 14(3):82-86.

Tonyaev, V.I. 1972. Geografia vnitrennikh vodnikh putey SSSR (Geography of inland waters of the USSR). Izd. Transport, Moscow. 176 pp.

Tretyakov, P. 1869. Turukhanskii District. Zapiski Imperatorskogo Russkogo Obshchestva po Obshchei Geografli 2: 215-530.

Trusov, V.Z. 1964. Some peculiarities of maturation and the scale of sexual glands maturity in sturgeon. Pages 69-78 in Trudy of All-Union Research Institute for Sea Fisheries and Oceanography (VNIRO) vol. 56 no. 3.

Tsepkin, E.A. 1995. Changes in the harvested fish fauna of continental water bodies in Eastern Europe and North Asia in the Quaternary Period. Voprosy Ikhtiologii (Issues in Ichtyology) 35(1):3-18.

Tugarina, P.Ya, M.G. Askhaev and E.S. Gomenyuk.

1967. Ecology of leschya from Lake Ubinsk in the Irkutsk reservoir. Izv. Biol. Geogr. NII Irkutsk Gos. Univ., Irkutsk. XX: 187-200.

Tugarina, P.Ya and E.S. Gomenyuk. 1967. On ecologic and biologic characteristics of fishes in Irkutskoe reservoir. Izv. Biol. Geogr. NII Irkutsk Gos. Univ., Irkutsk. XX: 201-253.

Tyaptirganov, M.M. 1988. Antropogennaya suksessiya vodnoi ekosistemy reky Khromy (Anthropogenic succession in the aquatic evosystem of the Khroma River). Institut. Biol. Ya.F SO AN SSSR, Irkutsk. 94 pp.

Urban, V.V. 1949. Hydrobiological studies of the mouth of the Lena River. Izv. Vsesoyuznogo Nauchno-Issledovatelskogo Instituta Ozernogo i Rechnogo Rybnogo Khozyaistva 29: 75-95.

Ushakov, B.P. 1963. Problems of Animal Cytoecology. Pages 5-20 *in* Sbornik Rabot no. 6 (Collection of Papers no. 6). Akad. Nauk SSSR, Moscow.

Usinin, V.F. 1978. Biology of the sterlet (*Acipenser ruthenus* L.) of the Chulym River. Voprosy Ikhtiologii (Issues in Ichtyology) 18(4):624-635.

Varpakhovskii N.A. 1897. Data on the fisheries in Eastern Siberia. Russkoye sudokhodstvo torgovoye i promyslovoye na rekakh, ozerakh i moryakh (Russian shipping in trade and fisheries on rivers, lakes and seas) 195: 69-84.

Varpakhovskii N.A. 1897. Data on the ichthyological fauna in the Ob River basin Part I. Saint Petersburg. 31 pp.

Varpakhovskii N.A. 1902. Fisheries of the Ob River basin. Pages 145-230 *in* Ryby reki Obi II (Fish of the Ob River Part II). Saint Petersburg.

Vasiliev, V.S., G.S. Ivan'kova, L.M. Maslov, M.A. Popov et al. 1957. Sanitarnaya kharakteristika reki Irtysha v raione g. Omska po dannym fiziko-khimicheskikh, bakteriologicheskikh i biologicheskikh issledovaniy. (Sanitary characteristics of the Irtysh river around Omsk according to data from physico-chemical, bacteriological and biological investigations). Izd. Minzdrav RSFSR. 147 pp.

Vasiliev, V.P., L.I. Sokolov and E.V. Serebryakova. 1980. Karyotype of the Siberian Sturgeon *Acipenser baerii* Brandt of the Lena River and Some Problems of Karyotype Evolution in Acipenseriformes, Voprosy Ikhtiologii (Issues in Ichtyology)., 1980, 20: 814-822.

Vedenov, M.F. and V.I. Kremyanski. 1967. Problems of the Interrelationships between the Structural Levels of Biological Systems. Pages 252-263 *in* Strukturnye urovni biosistem (Structural Levels of Biosystems). Akad. Nauk SSSR, Moscow.

Velichko A.A., Yu. M. Kononov and M.A. Faustova. 1994. The last glaciations of the pleistocene. Priroda 7: 63-67.

Venglinskii, D.L. 1974. Ecological aspects of rational harvest of coregonids in northwestern Siberia and the Ural. Pages 17-21 *in* Proceedings of the VI Symposium on Issues of the North, Yakutsk. Yakutsk fil SO AN SSSR.

Venglinskii, D.L. 1984. The impact of harvest on fish populations in northern Siberia. Rybnoye Khozaistvo (Fisheries) 2: 35-36.

Vershinin H.V. 1964. Characteristics of formation of benthic fauna in the lower Lena and its adventitious system. Tr. Sib. Otd. Vsesoyuznogo Nauchno-Issledovatelskogo Instituta Ozernogo i Rechnogo Rybnogo Khozaistva 8: 251-276.

Veshchev, P.V. 1979. Biological characteristics of sturgeon spawners at the spawning grounds of the Volga River. Pages 115-122 *in*

Biologicheskiye osnovy razvitiya osetrogogo khozaistva v vodoyemakh SSSR (Biological foundations for the development of sturgeon husbandry in the USSR). Nauka, Moscow.

Veshchev, P.V. 1991. Efficiency of the Natural Reproduction of the Stellate Sturgeon *Acipenser stellatus* in the Regulated Conditions of the Volga River Discharge, Voprosy Ikhtiologii (Issues in Ichtyology) 31: 222-227.

Vladimirov, V.I. 1957. On the Biological Classification of Fishes as Anadromous and Semi-Anadromous. Zool. Zh. 36: 1121-1125.

Voronina, E.A. 1974. Influence of Incorporated Radioactive Strontium on the Tilapia Male Gonads. Trudy Vsesoyuznogo Nauchno-Issledovatelskogo Instituta Ozernogo i Rechnogo Rybnogo Khozyaistva. 100: 84-92.

Voronina, E.A., Peshkov, S.P. and Shekhanova, I.A. 1974. Growth Rate and Fecundity in Fish Living in an Environment with Increased Radiation. Trudy Vsesoyuznogo Nauchno-Issledovatelskogo Instituta Ozernogo i Rechnogo Rybnogo Khozyaistva 100: 74-79.

Voroshilova, A. A. and E.V. Dianova. 1960. On the bacterial acidification of oil and its proliferation in natural water bodies. Mikrobiologia 19: 203-210.

Votinov, N.P. 1958. Acipenserids of the Ob basin. Tyumen. 43 pp.

Votinov, N.P 1963. Biological foundations of artificial reproduction of the Ob River sturgeon. Trudy Ob-Tazovskogo Otdeleniya Gosudarstvennogo Nauchno-Issledovatelskogo Instituta Ozernogo i Rechnogo Rybnogo Khozyaistva, Novaya Seria 3: 5-102.

Votinov, N.P. 1965. Conditions of cultivation of Siberian sturgeon in the reservoirs of the Upper Irtysh. Pages 109-112 in Teoreticheskiye osnovi pybkhozaistvo (Theoretical foundations of aquaculture). Moscow.

Votinov, N.P. 1966. Population dynamics of the sturgeon in the Ob-Irtysh basin and the impact of hydroconstruction and pollution. Pages 96-98 in Proceedings of the III Conference of Siberian Zoologists. Izd. Tomsk. gos. Univ., Tomsk.

Votinov, N. P, V. N. Zlokazov, V. P Kasyanov & R. I. Setsko. 1975. Sostoyanie zapasov osetra v vodoemakh Sibiri i meropriyatiya po ikh uvelicheniyu (Status of sturgeon reserves in the rivers of Siberia and measures aimed to increase these reserves). Sredneuralskoe Knizhnoe Izdatelstvo, Sverdlovsk. 94 pp.

Votintsev, K.K. 1961. Hydrochemistry of the Baikal Lake. Moscow. AN USSR. 311 pp.

Vrochinskii K.K and G. V.Semkov. 1978. Physical and biochemical data on the experimental intoxication of fish with yalan. Voprosy Ikhtiologii (Issues in Ichtyology) 18: 1128-1133.

Yablokov, A.V. 1996. Chernobyl and the environment. Pages 7-8 in Posledstviya Chernobyskoy katastrofy: zdorovye sredy (The consequences of the Chernobyl disaster and the environment). Zentr. Ekol. Polit, Moscow division of the international BioTest foundation, Moscow.

Yablokov, A.V. and A.G. Yusufov, A.G. 1981. Evolyutsionnoe uchenie (Evolutionary Theory). Vysshaya Shkola, Moscow. 344 pp.

Yakovlev, V.N. 1977. Phylogenesis of acipenserids. Pages 116-144 in Ocherki filogenii i sistematika iskopayemikh ryb i bezchelyustnykh. Nauka, Moscow.

Yakovleva, I. V, 1954. Development of Teeth in Sturgeon in Relation to the Stages of the Larval Development. Dokl Akad. Nauk SSSR

34(4):775-778.

Yarzhombek, A.A. and E.N. Bekina, E.N. 1987. The Efficiency of Extraction of Dissolved Substances from the Water by Fish. Voprosy Ikhtiologii (Issues in Ichtyology) 27(4): 658-664.

Yarzhombek, A.A., A.E. Mikulin and A.N. Zhdanova, A.N. 1991. The Toxicity of Substances on Fish, Depending on the Impact Mode. Voprosy Ikhtiologii (Issues in Ichtyology) 31: 396-502.

Yerbaeva, E.A. 1967. Dynamics of the colonization of the Irkutsk reservoir by representatives of the genus *Tendipes* Meig (*Diptera, Tendipedidae*). Izv. Biol.-Geogr Nauchno-Issled Inst. Irkutsk Gos. Univ. XX: 95-109.

Yerbaeva, E.A. 1967. Dynamics of the colonization of the Irkutsk reservoir by representatives of the genus *Tendipes* Meig (*Diptera, Tendipedidae*). Izv.Biol.-Geogr Nauchno-Issled Inst.Irkutsk Gos.Univ.XX:95-109.

Yerbaeva, E.A. 1973. Chironomids of the Angara River and its reservoir. Pages 179-180 *in* Water-bodies of Siberia and perspectives for its exploitation for fisheries. Izdatelstvo Tomskoga Gosudarstvenogo Universiteta, Tomsk.

Yereschenko, V.I. 1966a. Condition of fisheries in the reservoirs of the Upper Irtysh and measures for its development. Pages 280-283 *in* Biologican foundations for fisheries in the water ways of Central Asia and Kazakhstan. Izd. Nauchnogo Soveta Akademii Nauk SSSR, Alma-Ata.

Yereschenko, V.I. 1966b. The effect of hydroconstruction of Siberian sturgeon reproduction in the Upper Irtysh. Pages 67-69 *in* Proceedings on the biological productivity of Siberian water bodies (October 1966, Irkutsk). Short Reports LIN SO AN SSST, Irkutsk.

Yereschenko, V.I. 1970. Condition of Siberian sturgeon populations in the reservoirs of the Upper Irtysh and possibilities of its cultivation. Pages 158-163 *in* Acipenserids of the USSR and their cultivation. Nauka, Moscow.

Yesipov, V.K. 1923. Notes on the commercial fisheries of the Lower Lena. Rybnoe Khozaistvo IV: 28-49.

Yevtushenko, N.Yu., A.S. Potrokhov and O.G. Zinkovskii. 1991. The Impact of Heavy Metals on the Reproductive Functions of Carp Spawners and Some Indexes of the Nuclein Acids Metabolism in Somatic Tissues and in the Gonads of Fish. Page 30 *in* proceedings of the All-Union Conf. on Fish Reproductive Physiology, Minsk.

Yudanov, I.G. 1935. Perspectives on sturgeon fishing in the Ob Bay and Cape Noviy Port. Trudy Ob-Tazovskogo Otdeleniya Gosudarstvennogo Nauchno-Issledovatelskogo Instituta Ozernogo i Rechnogo Rybnogo Khozyaistva, Novaya Seria 3:7-12.

Yukhneva, V.S. 1970. Data on the biocenosis of the Ob Delta and their patterns of distribution. Pages 189-191 *in* Produktivnost' biotsenoza v subarktike. UNTs URO AN SSSR, Sverdlovsk.

Zakharov, V. M. and G. M. Clarke. 1993. Biotest: a new integrated biologiocal approach for assessing the condition of natural environments. International Biotest Foundation, Russian Affiliate, Russian Academy of Sciences, Moscow, Russia. 52 pp.

Zakharov, V.M. and E.Yu. Krysanov. Methodology of an environmental health index. Ecological condition of the Chapaevka River basin under anthropogenic influence (Biological indicators). Pages 226-232 *in* The Ecological safety and sustainable development in the Samarsk Oblast. Izd. Instituta ekologii volzhskogo basseina RAN, Toliatti, Russia.

Zakharova, N.I. 1983. Development of the Gonads of Rainbow Trout Following Radiation Impact; Report 1: Irradiation of 24 Days Old Larvae. Vopr Ikhtiol. 23: 951-960.

Zambriborshch, F.S. and Bu Lai. 1976. The Impact of Hexachloran and Chlorofose on Yearlings of the Leaping Gray Mullet *Mugil saliens* Risso. Voprosy Ikhtiologii (Issues in Ichtyology) 16: 930-936.

Zavarzin, A.A. 1938. Kurs gistologii i mikroskopicheskoi anatomii (The Course of Histology and Microscopic Anatomy), Medgiz, Leningrad. 631 pp.

Zavarzin, A.A. and S.L. Shchelkunov. 1954. Rukovodstvo po gistologii (Manual on Histology), Medgiz: Leningrad. 700 pp.

Zemkov, G.V. and G.F. Zhuravleva. 1987. Morphofunctional Assessment of the Detoxicating Function of the Liver and Reproductive System in Fish During their Spawning Migrations (with Reference to Sturgeons). Page 41 *in* I Vsesoyuzn. simpozium po metodam ikhtiotoksikologicheskikh issledovanii (First All-Union Symp. on Methods of Ichthyo-Toxicological Research), Leningrad.

Zimakov, I.E. and L.P. Kuznetsova, L.P. 1987. A Method of Evaluating Chemical Pollutants' Impact Based on Fish Reproductive Ability. Pages 42-43 *in* I Vsesoyuzn. sintpozium po metodam ikhtioroksikologicheskikh issledovanii (First All-Union Symp. on Methods of Ichthyo-Toxicological Research), Leningrad.

Zhuravleva, G.F., A.A. Romanov and G.V. Zemkov. 1991. The Impact of Ecological Factors on the Reproduction of Sturgeon, Page 31 *in* Proc. of Vsesoyuzn. soveshchanie: Reproduktivnaya fiziologiya ryb (All-Union Conf. on Fish Reproductive Physiology), Minsk.

Zubkova, E.I., A.M. Zelenin, and S.A. Biryukova. 1989. Pecularities of Microelement Accumulations in Goldfish from the Reservoirs of Moldavia. Pages 155-157 *in* proc. of VII Vsesoyuzn. konferentsiya: Ekologicheskaya fiziologiya i biokhimiya ryb (VII All-Union Conf. on Fish Ecological Physiology and Biochemistry), Yaroslavl.

Publications in other languages

Akimova, N. V. and G.I. Ruban. 1995. Disturbances of Siberian sturgeon's generative system resulted from anthropogenic influence. Pages 74-79 *in* Proc. of Internal. Simp. on Sturgeons (September 6-11, 1993. Moscow) VNIRO Publishing, Moscow.

Ansari Badre Alum, Kumar Kaushar. 1987. Malathion toxicity: effect on the ovary of the zebra fish Brachydanio rerio (Cyprinidae). Int. rev. gesamt. Hydrobiol. 72(4): 517-528.

Artyukhin, E.N. 1995. On biogeography and relationships within genus Acipenser. Sturgeon Quarterly. 3(2): 6-8.

Barannikova, L.A. 1993. Measures to maintain sturgeon fishery under conditions of changes in ecosystems. Pages 33-34 *in* Intern. Sump. on Sturgeons, September 6-11 1993, Moscow-Kostroma-Moscow. VNIRO, Moscow.

Barannikova, L.A. 1993. Measures to maintain sturgeon fishery under conditions of changes in ecosystems. Pages 131-136 *in* Intern. Sump. on Sturgeons, September 6-11 1993, Moscow-

References

Kostroma-Moscow. VNIRO, Moscow.

Beach, A.W. 1959. Seasonal changes in the cytology of the ovary and of the pituitary gland of the goldfish. Can. J. Zool. 37:615.

Bemis, W., E. Findeis and L. Grande. 1997. An overview of Acipenseriformes. Environ. Biol. Fish. 48: 25-71.

Berg, L.S. 1904. Zur Systematik der Acipenseriden. Zool. Anz. XXI(22):665-667. [In German].

Berg, B. and R. Hurk. 1983. Oocytes in the testes of the three-spined stikleback *Gasterosteus aculeatus*. Copeia. 1:259-261.

Birstein, V.J., R. Hanner, R. DeSalle. 1997. Phylogeny of the Acipenseriformes: Cytogenetic and Molecular approaches. Environ. Biol. Fish. 48: 127-155.

Birstein, V.J. and R. DeSalle. 1998. Molecular phylogeny of Acipenseridae. Molecular phylogenetics and evolution. 5(1): 141-155.

Brandt, J. F. 1870. Einige Worte über die europaisch-asiatischen Störarten (Sturionides), von Johann Friedrich Brandt (Lu Ic 20 mai 1869) Bulletin de L'Academie Imperiale des Sciences de St. Petersbourg. XIV: 171-175.

Brown, M.E. 1946. The growth of brown trout (Salmo trutta Linn.): Factors influencing the growth of trout fry J. Exp. Biol. 22(3-4): 118-129.

Bunge, A. 1883. Naturhistorische Nachrichten aus der Polarstation an der Lena Mündung. Bulletin de L'Academie Imperials des Sciences De St.-Petersbourg. XXVIII(4): 517-546.

Chew, R.L. 1973. The failure of largemouth bass, *Micropterus salmoides floridanus* (Losueur), to spawn in entrophic, overcrowded environments. Pages 306-319 *in* Proc. 26th Annu Conf. Southeast. Assoc.Game and Fish Commis, Knoxville. Tenn.

Crespo, S. 1990. L'Histopathologie en toxicologie aquatique. Bull. Soc. zool. Fr. V 115(2): 47-53.

Dumeril, A.H.A. 1870. Histoire Naturelle des Poissons, ou Ichthyologie generale. Vol. 2. Paris.

Findeis, E.K. 1993. Osteology of the North American shovelnose sturgeon *Scaphirhynchus platorynchus* Rafinesque 1820, with comparisons to other Acipenseridae and and Acipenseriformes. Univ. Of Massachusets, Amherst. (Ph. D. Thesis). 444 pp.

Findeis, E.K. 1997. Osteology and phylogenetc relationships of recent sturgeons. Environ. Biol. Fish. 48: 73-126.

Gardiner, B.G. 1984. Sturgeon as living fossils. Pages 148-152 *in* Living fossils. New York.

Georgi, I.G. 1775. Bemerkungen einer Reise im Russishen Reiche im Jahre 1772. St. Petersburg. 506 pp.

Grande, L. and W. Bemis. 1996. Interrelationships of Acipenseriformes, with comments on "Chondrostei". Pages 85-115 *in* (M.L.J.Stiassny, LR.Parenti, and G.D.Jonson, Eds.) Interrelationsips of Fishes, Academic Press, New York.

Gray, J. 1928. The growth of Fish II. The growth rate of the embryo of *Salmo fario*. J. Exp. Biol. 6(2): 110-124.

James, F.C. and C.E. McCulloch. 1990. Multivariate analysis in ecology and systematics: panacea or Pandora's box? Annu: Rev. Ecol. Syst. 21: 129-166.

Jin, F. 1995. Late Mesozoic acipenseriforttu (*Osteichthyes: Actinopterigii*) in Central Asia and their biogeographical implications. Pages 15-22 *in* Sixth Simposium on Mesozoic Terrestrial Ecosystems and Biota, Short Papers, (A.Sun, and Y.Wang, Eds.). Cina Ocean Press, Beijing.

Jin, F., Y. Tian, S. Deng. 1995. An early fossil

sturgeon (*Acipenseriformes, Peipiaosteidae*) from Fegning of Hebei, China. Vert. Palasiatica 33: 1-16 (in Chinese with English summary).

Lamarque, P. 1979. *Acipenser baeri en elevage a Donzacg (Landes).* Bull. Cent. d'etud et rech. sci. Biarritz 12(3): 545.

Luksiene, D. and O. Sandstrom. 1994. Reproductive disturbance in a roach (*Rutilus rutilus*) population affected by cooling water discharge. Journal of Fish Biology. 45: 613-625.

Mayden, R. L. and B.R. Kuhajda. 1996. Systematics, taxonomy, and conservation status of the endangered Alabama sturgeon, *Scaphirhynchus suttukusi* Williams and Clemmer (*Actinoptergii, Acipenseridae*). Copeia 241-275.

Mosimann, J.E. 1970. Size allometry: size and shape variables with characterizations of lognormal and generalized gamma distributions. J. Am. Statist. Assoc. 65: 930-945.

Narayan, Ram Raj and A.G.Sathyanesan. 1986. Ammonium sulfate induced nuclear changes in the oocyte of the fish *Channa punctatus* (Bl.). Bull. Environ. Contain. and Toxicol. 1990.:871-875.

Okada, Yo. K. 1965a. Bisexuality in sparid fishes 1. Origin of bisexual gonads in *Mylio macrocephalus*. Proc. Japan Acad. 41(4): 294-299.

Okada, Yo K. 1965b. Bisexuality in sparid fishes II. Sex segregation in *Mylio macrocephalus*. Proc. Japan Acad. 41(4): 300-304.

Pallas, P.S. 1814. Zoographia rosso-asiatica, sistems omnium animalium in extenso Imperio rossico et adjacentibus maribus observatiorum recensionem, domicilia mores et descriptiones, anatomen atque icons plurimorum. T.III. Petropoli. St. Petersbourg. 428 pp.

Pavlov, D.S., Ruban, G.I., Sokolov, L.I. 2001. On the types of spawning migrations in sturgeon fishes (Acipenseriformes) of the world fauna. Journal of Ichthyology, vol. 41, Suppl. 2, pp. 225-236.

Ruban, G.I. 1989. Clinal variation of Morphological Charactres in the Siberian Sturgeon, *Acipenser baeri,* of the Lena Basin. Journal of ichthyology (formerly Problems of Ichthyology). 29(7): 48-55.

Ruban, G.I. 1992. Plasticity of development in natural and experimental populations of siberian sturgeon *Acipenser baeri* Brandt. Acta Zoologica Fennica. 191: 43-46.

Ruban, G.I. 1994. Species structure, modern distribution and state of populations of the Siberian Sturgeon *Acipenser baerii* Brandt. The International Conference on Sturgeon Biodiversity and Conservation (The American Museum of Natural History, July 28-30, 1994 New York). Summaries of Oral Presentations. 16 pp.

Ruban, G. I. 1996. The Siberian Sturgeon, *Acipenser baerii baerii,* Population Status in the Ob River. The Sturgeon Quartrly. 4(1/2): 8-10.

Ruban, G.I. 1997. Species Structure, Contemporary Distribution and status of Siberian Sturgeon, *Acipenser baeri*. Pages 221-230 in (V.Birstein, J.R.Waldman, and W.E.Bemis eds.) Sturgeon Biodiversity and Conservation. Kluver Academic Publishers, Dordrecht.

Sauer, Martin. 1802. An account of a geographical and astronomical extpedition of the northern parts of Russia, by commodore Loseph Billings in the years 1785 to 1794. London. 332 pp.

Scott, D.P. 1962. Effect of food quantity on fecundity of rainbow trout, *Salmo gairdneri*. J.

Fish. Res. Board Canada. 19(4).

Sehgal, Rekha and A.B. Saxena. 1986. Toxicity of zinc to a viviparous fish, *Lebistes reticulatus* (Peters). Bull. Environ. Contam. and Toxicol. 36(6): 888-894.

Sokolov, L.I., V.P. Vasil'ev. 1989. *Acipenser baeri* Brandt, 1869. Pages 263-284 *in* The Freshwater Fishes of Europe. Vol. I/II General Introduction to Fishes Acipenseriformes. AULA-Verlag, Wiesbaden.

Suter, G.W., A.E. Rosen, E. Linder and D.F. Parkhurst. 1987. Endpoints for responses of fish to chronic toxic exposures. Environ. Toxicol. and Chem. 6(10): 793-809.

Waddington, C.H. 1957. The srategy of genes. Allen & Unwin, London. 262 pp.

Woodward D.M., R.G. Riley R.G. and C.E. Smith. Accumulation, sublethal and safe concentration of a refined oil as evaluated with Cutthroat Trout. Arch. Environ. Contain.